Mattering Press

Mattering Press is an academic-led Open Access publisher that operates on a not-for-profit basis as a UK registered charity. It is committed to developing new publishing models that can widen the constituency of academic knowledge and provide authors with significant levels of support and feedback. All books are available to download for free or to purchase as hard copies. More at matteringpress.org.

The Press' work has been supported by: Centre for Invention and Social Process (Goldsmiths, University of London), European Association for the Study of Science and Technology, Hybrid Publishing Lab, infostreams, Institute for Social Futures (Lancaster University), Open Humanities Press, and Tetragon.

Making this book

Mattering Press is keen to render more visible the unseen processes that go into the production of books. We would like to thank Joe Deville, who acted as the Press' coordinating editor for this book, the two reviewers Susanne Bauer and Thomas Binder, Delaina Haslam, for the copy editing, and Hyperkit for the design of this book.

Energy Babble

Andy Boucher, Bill Gaver, Tobie Kerridge,
Mike Michael, Liliana Ovalle, Matthew
Plummer-Fernandez, and Alex Wilkie

MATTERING PRESS

ISBN: 978-0-9955277-2-0 (pbk)
ISBN: 978-0-9955277-3-7 (ebk)

This publication has been supported by the ESRC's Energy and Communities
Collaborative Venture, grant reference RES-628-25-0043

Contents

Preface
Bill Gaver

This book is the story of how we designed and made a set of computational devices called Energy Babbles, and gave them to groups concerned with energy conservation to try out in their everyday lives.

Energy Babbles are like automated talk radios obsessed with energy. Synthesised voices, punctuated by occasional jingles, recount energy policy announcements, remarks about energy conservation made on social media, information about current energy demand and production, and comments entered by Babble users. Developed for members of UK community groups working to promote sustainable energy practices, the Energy Babbles are designed to reflect the complex discourses such groups navigate, as well as to provide information and encourage communication.

Manifesting as unusual-looking computational research devices designed for homes or public spaces, the Energy Babbles are networked to a behind-the-scenes server that collects and curates content from a variety of sources ranging from Twitter™ to the National Grid. We deployed them for several months to the communities, spurring a variety of reactions and stories that give new insight into the complex territory of community energy reduction.

The Energy Babbles are the product of a collaboration between designers and science and technology studies (STS) researchers, building on the Interaction Research Studio's long history of working with sociologists, but marking a deeper engagement with concepts drawn from STS. We had several goals: to experiment with design's interventionist possibilities for STS; to test STS's potential in providing new articulations and perspectives to design; and to see how together we might offer new approaches to environmental work.

Most of the time during the project it didn't feel like we were 'doing STS'. Instead, the project largely proceeded like most design research projects we do. Our day-to-day conversations were more likely to be about coiled cables and Raspberry Pis than assemblages, performativity, or the sociology of expectations. This process of embedding ourselves in a design project, one to which we all contributed, is reflected by the images that make up at least half of this book. Behind this activity, however, and certainly after it, ideas from STS shaped our design approach and ways of talking about the project. This is most explicitly captured in the written reflections that make up the other half of the book.

In trying to present the project, this mix of images and text seemed the best way to show how the collaboration between designers and STS researchers played out. But things turned out to be more complicated than that. Turning from our comradely design work to producing this book made it clear that, beyond a meeting of two disciplines, this was a collaboration between seven researchers, each with a unique configuration of expertise and interests, from product design to human computer interaction, from community engagement to working with bots, from STS to interaction design.

As our writing emerged, then, so too did our differences (as well as our disagreements). Rather than smoothing these over to produce a homogenous text, we have allowed our different views to co-exist. So, for instance, an essay about STS and speculative design is gently rebuked by another rejecting category labels for the design work shown here. A detailed description of composing musical jingles is given as much space as one considering design as a form of public engagement.

The result is a potentially disconcerting conversation between different voices, all engaged with the same (or related) topics, yet all addressing them from different points of view – rather like the Babble. The result is not a narrative delivered fully baked to the reader, but an accumulation of materials that invites, or even requires, readers' involvement. We think this is both reflective of the Babble, and – in preventing any one perspective from 'owning' the account – important for conveying the nature of the project to disciplinary audiences.

We hope the book isn't too much like the Babble, however. As we will discuss, the Babble was, often and purposely, frustrating for its users. In developing this book, we hope, in contrast, to inspire readers. We have tried to mix images and words to expose our collaboration, to intrigue sociologists with views into the contingencies of interdisciplinary work, designers with an example of how STS can provide new insight, technologists with a device that opens rather than solves problems, and environmental researchers with the reminder that there's more to energy reduction than saving energy.

Of course, in adopting the relatively open, 'uncooked' approach, we expect that readers will find their own interpretations of our work. And this, really, is what we hope for most of all.

About the team

The Energy Babble was produced through a collaboration between members of the Interaction Research Studio, a practice research group exploring computational systems for everyday life, and Mike Michael, a leading scholar of science and technology studies (STS). As colleagues at Goldsmiths, University of London, we had been orbiting around each other for some years. Funded by the Research Councils United Kingdom (RCUK) Energy Programme, the ECDC project offered us a unique opportunity to explore how social enquiry and design-led interventions could be tangled together to engage communities in new ways.

ECDC is our in-house name for a project we originally entitled 'Sustainability Invention and Energy-Demand Reduction: Co-Designing Communities and Practice'. Faced with such a ponderous name, we came up with the more informal 'Energy and Co-Designing Communities'. Not only was it easier to say and remember, but we liked the way it recalled a well-known rock and roll band.

The Babble/ECDC team: Bill Gaver, Tobie Kerridge, Mike Michael, Liliana Ovalle, Matthew Plummer-Fernandez and Alex Wilkie.

to engage communities from around the UK, ultimately with a novel device called the

Acknowledgements

Thanks to members of the practitioner groups who shared their perspectives and expertise, offered hospitality and adopted our research devices: the Geezers, Energize Hastings, Greening Goldsmiths, Low Carbon Living Ladock, Meadows Partnership Trust, Reepham Green Team, Sid Valley Energy Action Group, Transition New Cross.

We are grateful for the partnership and participation of academic and professional colleagues: Jimmy Aldridge, Andrew Dobson, Owen Dowsett, Fiona Fieber, Karen Henwood, Sabine Hielscher, Matthew Lipson, Sarah Marie Hall, Katherine Moline, Janine Morley, Rachel Murphy, Bridget Newbury, Martin O'Brien, Karen Parkhill, Nick Pidgeon, David Rose, Fiona Shirani, and Adrian Smith.

And special thanks to Goldsmiths colleagues: David Cameron, Jennifer Gabrys, Nadine Jarvis, Carole Keegan, Noortje Marres, Naho Matsuda, and Jen Molinera.

Energy Babble. Our account consists of a variety of materials — essays, descriptions, images

Introducing the project: Entangling speculation, design, and STS

Mike Michael

This book aims to present a specific example of collaboration between scholars and practitioners in design and science and technology studies (STS). In particular, it reports on the interdisciplinary project 'Sustainability Invention and Energy-Demand Reduction: Co-Designing Communities and Practice' (shortened to ECDC), the simple objective of which has been to examine how 'energy-demand reduction' is related to 'community'. As one project out of the seven funded under the Research Councils United Kingdom (RCUK) Energy Programme, the purpose of ECDC was to combine sociological and design methodologies to explore the parameters of 'energy-demand reduction'. In practice, this entailed the development and deployment of a series of empirical engagements of varying degrees of innovation: initial ethnographic visits to 'energy communities', probe workshops, the distribution of probe packs, a smart meter de-inscription workshop, workbook compilation, Twitter 'bot' design and deployment, experience prototype testing, the Energy Babble (the final designed artefact) deployment, and follow-up ethnographic visits.

Along the way, the project touched on many 'broader' issues, including: the local conduct of interdisciplinarity; the crossovers and contrasts between social scientific and design research; the purpose and practice of 'method', and the use and status of 'data'; the complex role of objects in the engagement with, and enactment of, 'publics', 'users', and 'citizens'; emerging speculative conceptualizations of 'social events'; the character and problems of batch production; and so on.

These broader issues reflected not only ECDC's trajectory, but also the longer-term trajectory of which ECDC was a part. ECDC after all emerged out of, and was grounded in, ongoing discussions around the intersections of STS and design at Goldsmiths, University of London. These discussions took numerous forms including collectively organised events (e.g. the Design and Social Science Seminar Series, Goldsmiths 2009 – continuing; the Making and Opening Conference, 2010; conference sessions at EASST, 2010 and 2012) and co-publications (Michael and Gaver 2009; Wilkie and Michael 2009; Wilkie et al. 2010). These co-productions have been instrumental in both opening up design research to the conceptual resources of social science (and especially STS, e.g. Latour 2005; Law 2004; Stengers 2005) and, conversely, refreshing social scientific thinking about method and the relationship of research to its 'objects of study' (e.g. Di Salv, 2012; Dunne and Raby 2013). On this score, in ECDC we can witness further elaborations of STS-and-design interdisciplinarity. Initial articulations can be found in a number of recent publications (Wilkie et al. 2015; Michael et al. 2015; Gaver et al. 2015). The present book pulls together many of the insights gained over the last six or seven years to fashion a novel analysis of energy-demand reduction. More precisely, in relation to the RCUK programme, and in the context of current thinking about energy-demand reduction, ECDC, draws on and serves as a critique of particular models of energy consumption, or practice, or political action. As such, it interrogates the processes by which 'community' is performed, the means by which the 'future' is projected, how 'energy' unfolds in the practices of energy community members, how 'information' and 'knowledge' are constituted and deployed, and the parts played by design and social scientific 'method' in enabling and enacting these interconnected processes.

Hopefully, these brief introductory remarks give an indication of the scope and aspirations of ECDC. Representing all of this in a single volume has not been without difficulty – not least because of the contrasting genres of writing in STS and design, the disparate audiences to which STS and design scholars address themselves, and, of course, the different points of reference that structure designers' and STS-ers' respective perspectives on and within ECDC. While ECDC proceeded more or less happily as an interdisciplinary collaboration, the tacit divergences that were practically negotiated and pragmatically displaced in the process of working together came to the fore when it was time to start 'writing up' the project. The format of this book has thus gone through a number of iterations – and we thank the publishers for their understanding in this regard. In the end, we collectively decided to try to 'enact' the synergies and tensions that mark interdisciplinary work (including disciplinary takes on what counts as interdisciplinarity) in the form of the text itself.

As a result, we have eschewed a 'straightforward' narrative, and have instead developed a format which

— that come together to tell the story of a project. To try to help you, dear reader hopefully captures the diversity of modes of practice and scholarship that have gone into ECDC. Of course, we are also aware that any format is also a performance – things are left out, 'Othered', and under- or over-emphasised. As many authors have noted (e.g. Law 2004), scholarship whether as method, practice, theory, or writing-up 'makes' the object that it engages. The point thus becomes one of opening up potential new ways of thinking about the 'object' of study, proposing new opportunities for action, and enabling speculation on the possibilities that inhere in particular events in the present case – broadly speaking, the events of energy-demand reduction.

In any case, we have designed this volume in three main sections and with six broad ingredients. The sections characterise the project in terms of an opening phase of 'Framing', in which we established and engaged with the complex setting of communities, policies, and technologies for our work; a phase of 'designing', in which we explored and later refined ideas of what we could make in response to – not necessarily for – this setting; and a final phase of 'Circulating', in which we took the project outwards, both to the communities for whom we designed the Babble, and also to a more complex set of publics ranging from politicians to school children, and specialist researchers to passers-by wandering into exhibitions. Of course, the boundaries between these sections are notional, as the modes of work they describe didn't just leak across them but re- and preverberated throughout the project. Nonetheless, they reflect both the balance of our activities at any given point, as well as what we thought we were doing – and may in any case help orient the reader.

As for the book's 'ingredients', they combine more or less continuously throughout the three sections. The first is a running storyline which presents an 'unadorned' account of what we did, when we did it, how we did it, and so on. It is an orienting device which ideally allows the reader to follow the trajectory of the project, despite the proviso that this is necessarily partial. The second is a series of anecdotes – short accounts of the experience of working on some aspect of the project. Again, we do not see anecdotes as transparent representations of the past – they are at once pro-active and re-active insofar as they both reflect (topologically), and are enabled by the past, but also actively constitute that past (Michael 2012). Anecdotes therefore remind us that we as researchers are emergent from the process of research even as we are engaged in 'making' it. We then have a series of essays – these are longer pieces that situate the ECDC project against the backdrop of such concerns as energy policy, sound design, the operation of a design studio, working as an interdisciplinary group, batch production, and running a probe workshop. The aim here is to relate ECDC to a range of practical conditions that have impacted on the project, and which the project has, in turn, 'appropriated' in various ways. Fourthly, there is a number of longer articles. These take a more usual academic form and discuss empirical, conceptual, and methodological issues that link the project with evolving debates in relevant fields such as public engagement, speculative methodology, design research, interdisciplinarity, and practice theory. Fifthly, as a way of enacting our interdisciplinarity, and demonstrating the sometimes-tangential perceptions of each other's work, here and there we have notated the anecdotes, essays, and articles, but in the present case the notes do not derive from the authors themselves but from colleagues within the project team. In particular, we aim to evoke how a different sense could be made of the texts. The final element is the visual matter – photographs of events and objects, blueprints of designs, and reproductions of probe data. While this is sometimes used illustratively to support some point being made in the text, at other points it is used to stand in contrast to the text, to act as a resource through which to trace a different path through the project.

References

DiSalvo, C., *Adversarial Design* (Cambridge, MA; London: MIT Press, 2012).

Dunne, A., and F. Raby, *Speculative Everything: Design, Fiction, and Social Dreaming* (Cambridge, MA: MIT Press, 2013).

Latour, B., *Reassembling the Social* (Oxford: Oxford University Press, 2005).

Law, J., *After Method: Mess in Social Science Research* (London: Routledge, 2004).

Michael, M., anecdote, in C. Lury, and N. Wakeford (eds), *Inventive Methods: The Happening of the Social* (London: Routledge, 2012), pp. 25-35.

Michael, M., B. Costello, I. Kerridge, and J. Mooney-Somers, 'Manifesto on Art, Design and Social Science – Method as Speculative Event', *Leonardo*, 48(2) (2015): 190-91.

Michael, M., and W. Gaver, 'Home Beyond Home: Dwelling with Threshold Devices', *Space and Culture*, 12 (2009): 359-70.

Stengers, I., 'The Cosmopolitical Proposal' in B. Latour, and P. Webel (eds), *Making Things Public* (Cambridge, MA: MIT Press, 2005), pp. 994-1003.

Wilkie, A., W. Gaver, D. Hemment, and G. Giannachi, 'Creative Assemblages: Organisation and Outputs of Practice-Led Research', *Leonardo*, 43(1) (2010): 98-99.

Wilkie, A., and M. Michael, 'Expectation and Mobilisation: Enacting Future Users', *Science Technology Human Values*, 34(4) (2009): 502-22.

Wilkie, A., M. Michael, and M. Plummer-Fernandez, 'Speculative Method and Twitter: Bots, Energy and Three Conceptual Characters', *The Sociological Review*, 63(1) (2015): 79-101.

to orient, we are including this stream of text, which runs at the bottom of every page, to help stitch together the narrative. To begin with, we explain how we set out

FRAMING

At the outset we aimed to establish a network of participants and an identity to foster involvement. A workshop that gathered together low carbon community advocates and university researchers offered an opportunity for us to present a poster about the aims of our research project. Following this event, a design agency created a print and online identity for our project, which we called Energy and Co-Designing Communities, or ECDC.

Practitioners we had met at the workshop were contacted, and with their help we visited the communities where they had undertaken energy-demand reduction measures, and learned about their projects, infrastructures, and future plans. A workshop was convened at a museum of the home, where reduction practitioners, researchers and policy makers shared stories about energy and mapped imaginary communities. Cultural probes were made and given to community members, and their responses were arranged on the wall of our studio.

We experimented with demand-reduction practices, installing smart energy monitors, insulating our lofts and using software to visualise our energy use. The quantification and comparison of energy data seemed inescapable, though ultimately we embarked upon a lively re-imagination of smart monitors.

A prehistory
Mike Michael

From a standing start, we wrote the proposal in about two weeks. To say the least, we were delighted and not a little disconcerted when we heard the news that we had been successful.

I hadn't had much to do with the design department at Goldsmiths, but was introduced to design through Alex Wilkie, and later Tobie Kerridge. Alex approached me as a prospective PhD student; Tobie asked me to get involved in what was to become the Material Beliefs project. It was through them that I became aware of, and then visited, the group at the Interaction Research Studio, at the time housed in the old hut complex. My first impressions were that there was never much going on, although clearly something was happening as the place was littered with bits of technology, diagrams, models, and finished

his enthusiasm and ability to discuss ideas from science and technology studies (STS

artefacts such as the Drift Table. I can't remember the first meeting, but I do recall meeting Bill Gaver for the first time and hitting it off with him pretty much immediately. The same went for the other members of the Studio … they all seemed such great people. No doubt this was helped by not infrequent visits to the pub.

If my personal encounter with the Studio was a happy and sociable one, my intellectual response to the work going on there was altogether more fraught. At base, I just didn't get it. For instance, on being told about, and subsequently reading up on, Tobie's project 'Biojewellery' (in which jaw bone tissue was cultured and combined with precious metals to produce jewellery), I responded (internally) with a mixture of confusion, frustration, and anger. This was supposed to be an exercise in something like Science-in-Public, and yet from my sociological/science and technological studies perspective it made no sense. Similarly, on hearing about the Drift Table, or the Plane Tracker, I was again struck by their strangeness – these objects simply didn't make much sense to me.

Having said that, they did make me laugh – they affected me as things that remade the world in intriguing ways. Over many discussions with Alex, Bill, and Tobie (and later Andy Boucher, but also Matt Ward), I began to see some sort of promise in these objects and the processes behind them, though I couldn't at the time specify what such a promise entailed other than some vague feelings that it might have implications for how social science gets done. In any case, in 2008, we decided to start up a joint seminar series – Design and Social Science – between the Centre for the Study of Invention and Social Process in the sociology department, and the Interaction Research Studio. I'm not sure we had much of an idea what we were doing other than thinking a seminar series would be an interesting means through which to explore the possible intersections

of design and social science. Certainly, I just wanted to find out more about design, and in particular the version of design practised in the Studio. In retrospect, this was as much about immersing myself in a design environment, absorbing some version of design's 'ethos', for want of a better term. This was not an easy process – while I gave various presentations about the relation between design and social science, these were (again on reflection) embarrassingly misconceived: even at the time I was aware that I was missing the mark by quite some margin even if I could neither pinpoint the mark nor measure the margin. At the same time, I was having great conversations with my newfound design colleagues.

It was during that time that Bill Gaver and I decided to develop a research proposal on how design and sociology might work together to develop technologies that in mediating the experience of nature and 'the sublime' might also complexify that experience (rather than dissipate it). This was not funded. We also co-wrote a paper on the ideas of home and dwelling. For me, this was a pivotal moment when I seemed to 'get' (at least to my own satisfaction) the design that was being practised in the studio. Over the writing of this paper, and Bill's gentle prompts (Michael and Gaver 2009), I got to see that the playfulness, ambiguity, and unpredictability of the Interaction Research Studio's 'threshold devices' could enable people to engage with the ambiguous, unpredictable, and complex flows – flows that were at once social, technological, and natural – that composed the home.

Needless to say, my changing appreciation of the Studio's work was also shaped by my reading at the time, especially of Whitehead and Stengers, but I suspect that it was also affected by a number of sociomaterial arrangements. The Studio had relocated from some rather dingy old campus huts to a wonderfully light, top-floor space

in the Ben Pimlott Building. Having the seminars and meetings there made a real difference to the mood – the atmosphere – of our conversations. The visits to the pub also continued to lubricate my fascination with design. And my continued interaction with the artefacts themselves gave me a 'feel' for what they did, sociologically speaking.

I can recall that by, I guess, late 2009 we had reached a point where we all felt that something 'properly collaborative' was on the cards. It was time to work together on a project. It was then that the Goldsmiths Research Office approached us. There was a colleague who wanted to develop a proposal for the not-so-recently advertised Research Councils UK programme on 'energy communities'. As it turned out, the colleague's interests didn't map onto the programme's parameters, but we saw this as an opportunity to develop a proposal that drew on our respective concerns. I don't think any of us expected to get funded. For my part, I thought we could use this as a way of thrashing out what a collaboration might look like and how we might fold in the interests of designers and sociologists around a pretty well-defined topic – energy-demand reduction. From a standing start, we wrote the proposal in about two weeks. To say the least, we were delighted and not a little disconcerted when we heard the news that we had been successful.

References
Michael, M., and W. Gaver, 'Home Beyond Home: Dwelling with Threshold Devices', *Space and Culture*, 12 (2009): 359-70.

Project identity

One of the first things we worked on was establishing a visual identity for the project. This reflected the public-facing nature of the project. The funding proposal was translated into visual material which included a poster for a meeting that marked the start of the project, and graphic elements for a website, stickers and note paper to provide a visual coherence for subsequent activity.

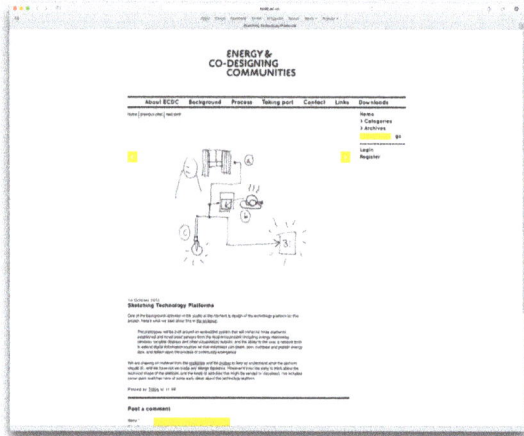

Layout of a Wordpress site for the project, an outlet for informal descriptions of events and processes

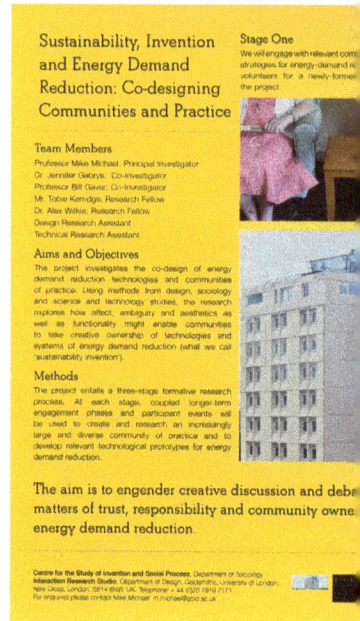

This poster describing the aims of our research project was presented at a workshop

Visual identity for the project, including logo and stationery

Fieldwork

From the outset, we inherited an inclination for the winners of the Low Carbon Communities Challenge (LCCC) from our funding body. These groups had made successful proposals for government grants to support a variety of energy-demand reduction measures in rural and urban settings across the UK.

Our approach was not to evaluate the effectiveness of the groups' activities, but to undertake fieldwork in order to understand energy-demand reduction practices, and to meet individuals who might participate more closely with the project and adopt the devices that would eventually be designed.

We contacted the groups we had spoken to at the initial workshop, identified and spoke to other funded communities, and cast our net beyond the LCCC groups while also looking much closer to home.

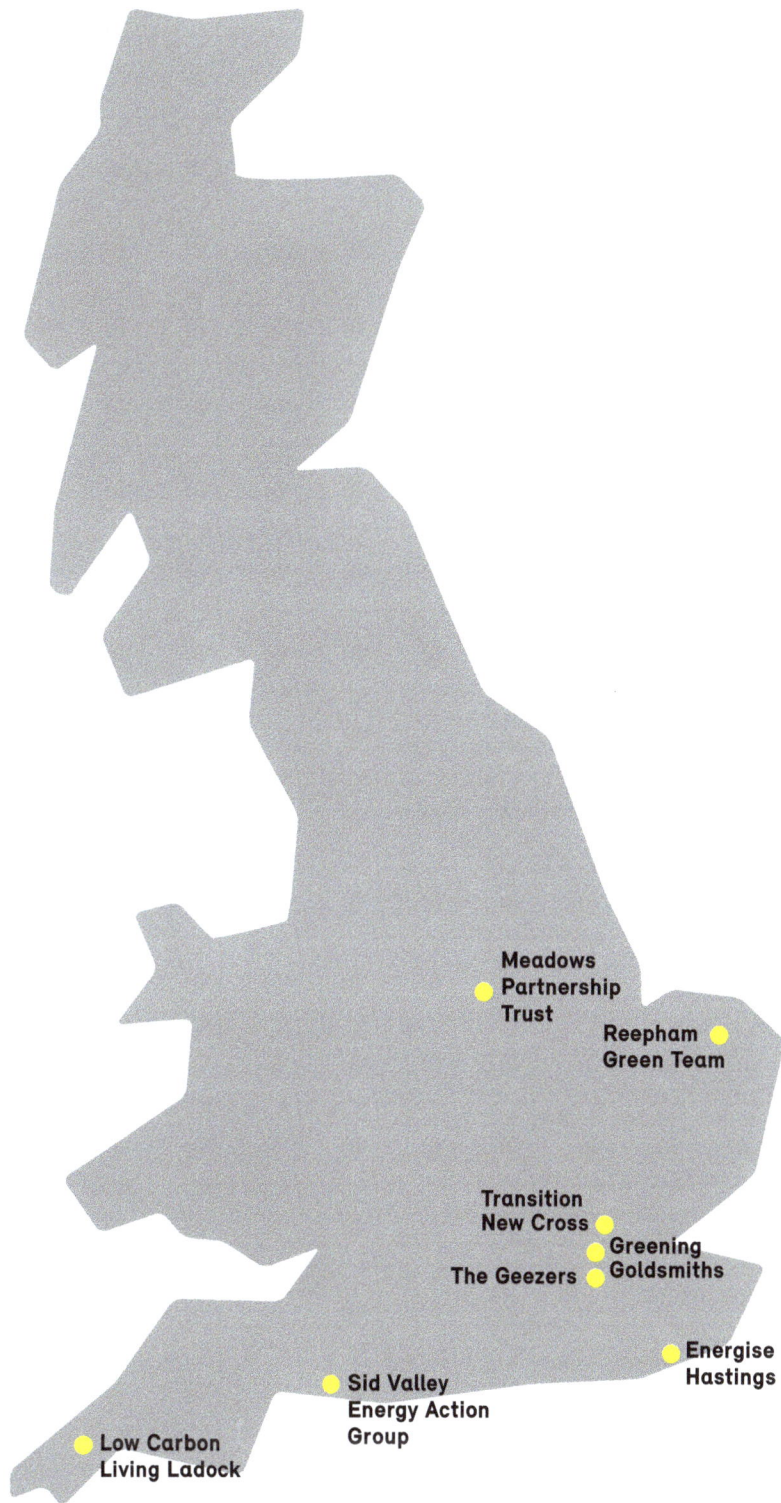

Meadows Partnership Trust

Reepham Green Team

Transition New Cross

Greening Goldsmiths

The Geezers

Energise Hastings

Sid Valley Energy Action Group

Low Carbon Living Ladock

Meadows Partnership Trust

The Trust is an organisation whose activities focus on the Meadows and Clifton area of Nottingham. In 2009 the Trust set up MOZES, an energy services company tasked with reducing the carbon footprint and energy costs of local residents, along with a variety of public engagement activities. MOZES made a series of successful funding proposals, including one to the Department of Energy & Climate Change LCCC competition.

After meeting Jacky, the manager of the Trust, at the kick-off meeting in London, we arranged a visit to the Meadows to hear more about these projects. We heard about a focus on energy efficiency measures for local housing, delivered through grants and interest-free grants from MOZES, to help people on low incomes to reduce the cost of their energy bills. We visited Arkwright Meadows Community Gardens, which produces organically grown fruit and vegetables, provides volunteering, education and training, and offers a tandoor oven for the community.

The Trust, and the suite of services provided through MOZES, is part of a rich and extensive network of services focused on the social and economic development of an urban area, and in this way offered a distinctive set of perspectives.

These were resident groups from different UK communities who had joined together to reduc

Reepham Green Team

A network of practitioners and local organisations initially came together to formulate a plan for the reduction of carbon emissions in the Norfolk market town of Reepham. We met Rex, who played a central role in the activities of this group, at the kick-off workshop. He arranged for us to visit Reepham where we heard more about the extensive portfolio of projects that the group had put in place following support from the LCCC competition.

Interventions included a mix of photovoltaics, roof insulation, wind turbines, pipe lagging, and double glazing at Reepham's primary school and sixth-form college; and at St Michael's Church a ground source heat pump provided an under-floor heating system. Elsewhere, LED street lighting

had been installed in the high street, and the town hall had been fitted with energy-saving light bulbs and efficient radiators. Additionally, eight bungalows had been refitted with triple-glazing, photovoltaic panels and air source heat pumps.

Our visit included three meetings, with representatives from Reepham Primary School and the High School as well as the Reepham Green Team, where we presented our project and heard how community members envisaged participating in the project. There is a clear sense of direction and responsibility within the group, who are delivering a broad set of services for Reepham. Indeed, there was some alignment here with the idea of big society, where it was seen that the group had acted to deliver services where local authority funding had been reduced.

Low Carbon Living Ladock

Ladock and Grampound Road is a rural parish near Truro in Corwall, with an active Transition group that undertook a series of sustainability measures following funding from the Department of Energy & Climate Change's LCCC competition. Chris, a core member of the group of about seventy-five residents, hosted us at Woodland Valley Farm, one of the sites where a range of renewable energy technologies was installed. He explained that the group sought to implement a 'framework for a world where less fossil fuels are used'. A mix of solar thermal and photovoltaic panels, and ground and air source heat pumps was installed at two schools, two community halls, two pubs, and eight homes.

The wind turbine mounted at the top of the valley acted as a 'visible symbol of local energy generation and carbon reduction'. The aim was not only that the turbine and other renewable energy measures would be evident to parish residents, but that it would operate for the shared benefit of the community.To support this ambition, the group had set up two companies, one to raise income from grants, donations, and government feed-in tariff payments, and the other to spend income on infrastructure that provided low-cost energy for the community.

The group worked with a range of actors in order to interpret and also inform policy, including parish and county councils, and national groups including the all-party parliamentary group on peak oil and gas. They also worked closely with Community Energy Plus, a low-carbon charity initially set up by Cornwall County Council, who supported their LCCC proposal.

Greening Goldsmiths

An internal Goldsmiths email asked staff to provide access to their workspaces so that roof insulation could be installed. The message was from Richard, who acted as Energy & Environmental Manager for the university. With colleagues, he ran Greening Goldsmiths, an initiative that plans and delivers a range of measures across the campus, to bring down energy demand and 'encourage thoughtful resource use'.

We spoke to Richard about his role, which included preparing tenders for solar panels to heat water, finding local fuel for an on-site wood pellet burner, and replenishing the stock of bees for the hives at the southern edge of the campus. Following a tour of Goldsmiths, and having visited the bees and the burner, we came across Richard amongst the rubble of a repair to a pipe that had rusted away, leaving a substantial building without hot water and heating. He had uncovered a map of the campus heating system in the library, and contractors had then managed to locate the breach.

As a consequence of being shown this infrastructure, and through experiencing its failure and renewal, our working environment became immanently tied to the activities of Greening Goldsmiths. In this way, energy communities became experienced not only through formal fieldwork activities, but habitually.

Transition New Cross

The Transition movement provided a set of core values for sustainable change that many of the groups we met were committed to. Consequently, when we aimed to identify a group local to our workplace in New Cross, we sought a Transition group. Our first meeting with members of this group was at the New Cross People's Library, a community-run space providing a range of services including a library and learning activities.

Transition New Cross was a loose and ad-hoc group of local residents that met irregularly at venues including the library, Green Shoots Community Garden, and the Hill Station community café on Telegraph Hill to discuss and deliver activities. We met a number of members, including Adrian, who was making a documentary film about sustainability and advocacy. Members also supported a range of related initiatives, including a co-operative food group.

In contrast to the other groups we worked with, Transition New Cross was not funded through large schemes, and in this respect was mobile and broadly networked. Membership and project activities were low-key and circumspect, though affiliations and commitments were deeply held.

up heat pumps, and using solar-powered showers. In the face of all this dedicated and

Energise Hastings

We initially approached Jane, a climate change project officer at Hastings Borough Council, who had set up Energise Hastings as a forum for individuals and organisations with an interest in renewable energy projects. This group held regular meetings at a range of local venues including the White Rock Hotel, and we attended these events as researchers and also became directly involved in the activities of the group through an interest in local regeneration.

Through meeting individuals and hearing about projects at these meetings, we visited renewable energy installations at three community centres supported by the Energy For Tomorrow fund from British Gas, and toured a renovation project at an arts venue in the town centre that included external wall insulation. We spoke to Richard who had ambitious plans for a community biogas plant to provide district heating for West Hastings.

Energise Hastings was also seeking advice on setting up a legal framework in order to be eligible for Government funding. As with other groups, we were struck by the personal commitment and enthusiasm of its members. Additionally, individual interests and roles were extremely diverse, sometimes founded on ecological concerns, at other times expressed through enthusiasm for design and technology, or otherwise driven by entrepreneurial or charitable motivations. The group was testament to the vitality of energy communities, and indeed to the complexities presented by this mix regarding the development of a shared strategy.

practical activity, it was challenging to explain what we were setting out to do. Yes,

The Geezers

We came across The Geezers through a project called ActiveEnergy, in which a group of male pensioners partnered with artists and researchers and aimed to install tidal turbines at the Thames Barrier. The project was part of a busy programme of community activates undertaken by this group based in Bow, East London.

After being introduced to Ray, a leading member of the group, by Loraine, their academic partner on ActiveEnergy, we were invited to visit The Geezers at their regular meet-ups hosted by Age UK. Here we heard more about their activities, which included an interest in understanding how new technology can be used to generate free energy.

Their dealings with renewables were on the one hand motivated by the high cost of their utility bills, but also supported their keen engagement with local organisations. University research groups provided resources and expertise to support their ideas; heritage sites provided venues for meetings as well as offering geographical access to tidal power; and initiatives undertaken by schools and arts groups linked The Geezers' projects to the activities of local communities.

It was evident that the elaborate and energetic partnerships that The Geezers had cultivated, for which the Age UK meetings acted as an unlikely hub, kept these men connected to the rapidly changing area of East London where they had spent their whole lives.

we would be building devices. No, they might not be immediately helpful to them. Instead,

Sid Valley Energy Action Group

The secretary of this Transition group, located in the East Devon town of Sidmouth, approached us after hearing about our project from a relative of one of our researchers. Sid Valley Energy Action Group (SVEAG) is part of the Vision Group for Sidmouth, formed more than ten years ago following a town council meeting discussing sustainability and renewable energy. We met Louise at the Anchor Inn, to take part in a monthly meeting of SVEAG.

Members of the group took turns to report on the outcomes of related activities, or to propose future events. Social events including a barbecue and an alternative energy vehicle show were seen as opportunities to build the culture of the group and also recruit new associates. Involvement with ECDC was understood to be an opportunity to develop the local 'impact' of their activities by potentially reaching a wider audience through local media coverage. One member reported on a visit to nearby Wadebridge, a Transition Town with a wide membership that was undertaking a variety of sustainability projects. This group was perceived as offering a successful model that Sid Valley might aspire to.

In common with other Transition groups, SVEAG believed that community action came about through the incremental development of an active and committed group of people. They anticipated the incorporation of an Industrial Provident Society as a key step, at which point the group would be eligible to apply to a range of funding opportunities. Capital would resource more ambitious projects, including an anaerobic digestion facility that used waste from tourism (an industry on which the prosperity of this market town was largely based) as fuel.

the designs would address the 'big picture' of the communities' situations and practices ...

A workshop

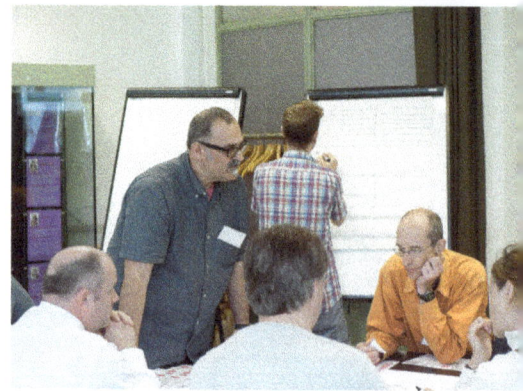

The interior of the Geffrye Museum in London is arranged as a series of historic domestic interiors, including period furniture and appliances. Considering the frequent focus on houses as the basic unit of intervention for renewables, this felt like a useful setting for a workshop for participants and researchers to come together and undertake a series of activities that would motivate and inform subsequent project activity.

The workshop began with a speed dating session, where participants had a chance to introduce themselves to each other, and hear something about one another's backgrounds. There followed a set of activities, including mapping an imaginary energy community, drawing domestic energy stories as floor plans, and writing hypothetical newspaper front pages. The day was concluded with the distribution of probe packs to attendees, to be taken away, used, and returned.

Attendees took part in a range of activities that aimed to generate unexpected and imaginative accounts of their practices and perspectives

To further explain our approach, and get to know the communities better, we hosted

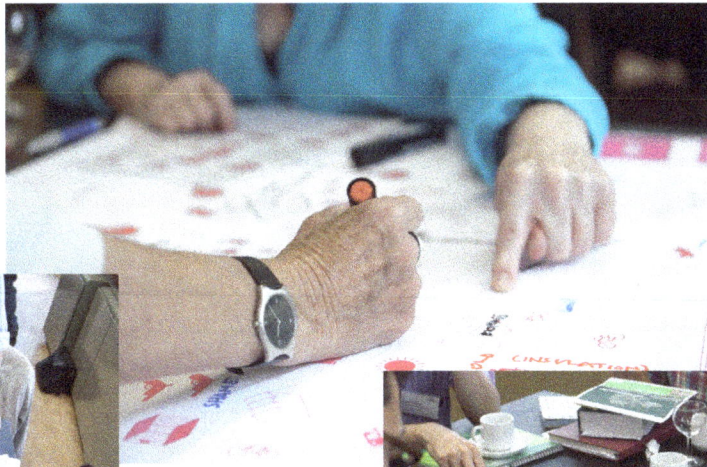

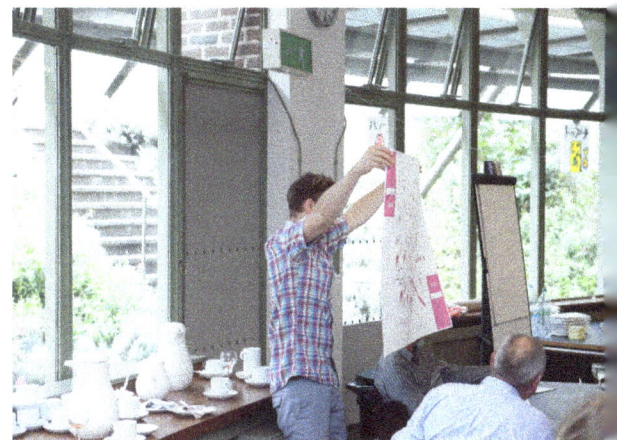

workshop in which we asked participants to engage with a series of playful tasks as a way

ROAST DINNER.

① EATING DINNER WITH FRIENDS

② PUTTING ON / UNTYING CHICKEN

③ COOKING

⑤ RELAXING / FEELING VERY FULL

Color Key:

STRESSED — RELIEVED

STRETCHING CHICKEN THING.

Floorplans and stamps were used to create diagrams that represented domestic energy events in relation to particular emotions

of opening discussions. For instance, we asked people to draw the trajectory of an event

involving energy in their home, had them produce overlaid maps showing their current and

The Broadsheet

Great Value Breaks
FROM £ 3500,000.

30 Framing

LAST SET
LEAVES
HEATHROW

LAST NATO
SOLDIER IN
AFGHANISTAN

INTERNATIONAL p.45

TheBroadsheet

Tuesday 12.07.20 20

UN declares WAR ON NIMBY'S

- KYOTO TARGETS MUST BE MET
- SCEPTICS SIDELINED

UK SHAMED - 2020 RE TARGETS MISSED.

GREECE STILL IN DEBT

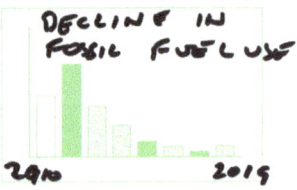

ECONOMICS p.13

DECLINE IN FOSIL FUEL USE

2910   2019

JULY-AUGUST 20 78
UNITED KINGDOM No.138
£ ___00

The Localist

12th July 2020   North Western Borough - Serving the local community since

p.5 BIOFUELS SUCCESS BOOSTS BRITAIN'S FARMERS

p.23 FIREMAN TEACHES FIRE SAFETY AT FETE.

Success with

GIANT PUMPKIN GROWN ON ORGANIC CERTIFIED WASTE NAPPIES!

COUPLE CELEBRATE CENTENARY WITH LOCAL MAYOR.

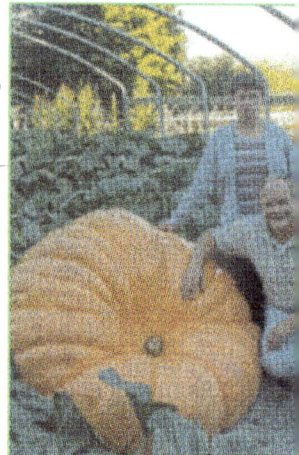

Classified
ELECTRIC CAR 50,000 MILES NEW BATTERIES £20,995
TOYOTA HYBRID 25K £12,995

Evening TV
HORIZON
WORLD ABOUT US
CIVILISATION
EQUINOX
NEWS AT 9
BACK TO THE FUTURE 5

Weather
WEDNESDAY CLOUDY SPELLS
THURSDAY AVERAGE
WARM, PLEASANT SUNSHINE 21C.

Architecture
ARCHITECTURAL MAGAZINE, DECORATION, ARTS, DESIGN

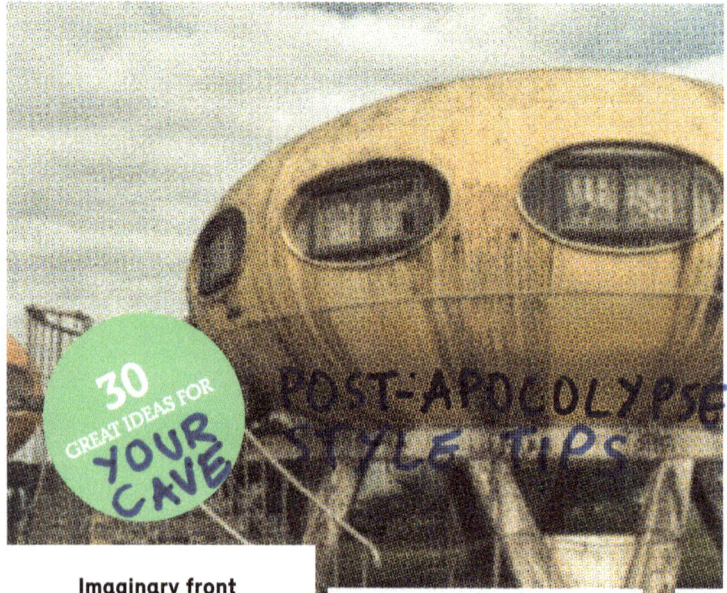

30 GREAT IDEAS FOR YOUR CAVE

POST-APOCOLYPSE STYLE TIPS

Imaginary front pages of magazines and newspapers spoke of sometimes dystopic and bizarre futures

The New Trends:
PUMPKIN HOUSES BECOME REALITY

WEEKLY | NEWS IDEAS INNOVATION   THE BEST JOBS IN SCIENCE

The Scientific

June 2050

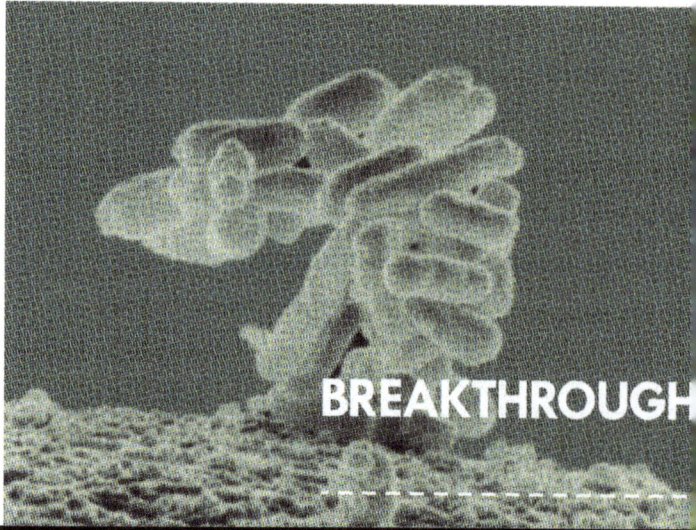

BREAKTHROUGH

imagined communities, and gave them incomplete newspapers and magazines to tell us the

THE *Tabloid*

Thursday, September 7, 20__61.   50p

FREE TODAY

GET YOUR GAS MASK

IT'S FULL !!!

CLIMATE REFUGEES CAMP IN HYDE PARK

energy news of the future. This gave us a sense of their concerns about energy and

Cultural Probes

Cultural Probes are a design-led approach to engaging with settings aimed at producing inspiration rather than information. They involve presenting people with open-ended, even absurd tasks in the hope that their responses will provide fragmentary insights into their lives, thoughts, hopes, and fears. Invented by Tony Dunne, Bill Gaver, and Elena Pacenti for a project spanning three European countries, they are often designed to rely on photographs and drawings, as well as short written responses, to minimise reliance on language and provide relatively direct glimpses into peoples' situations.

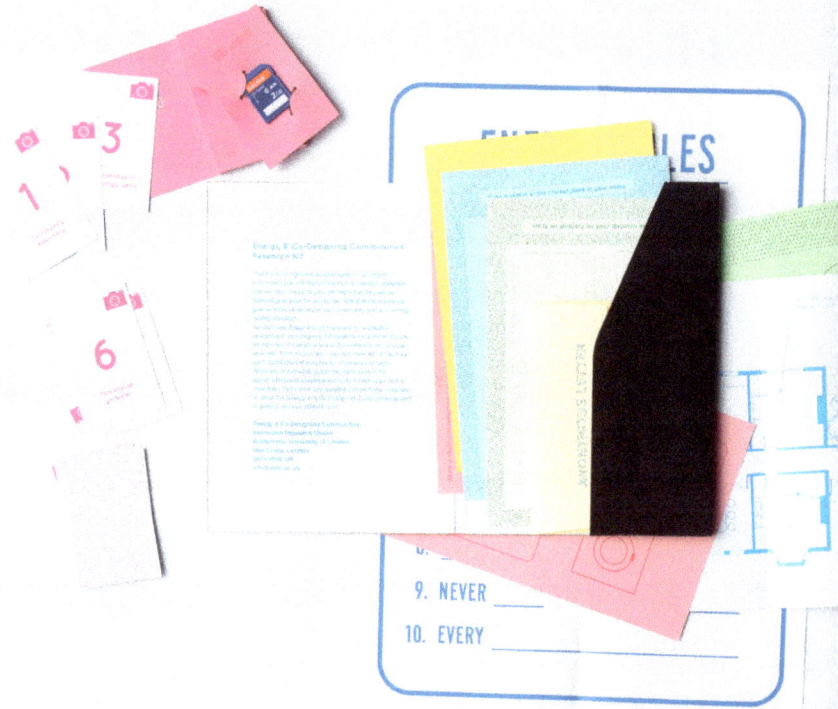

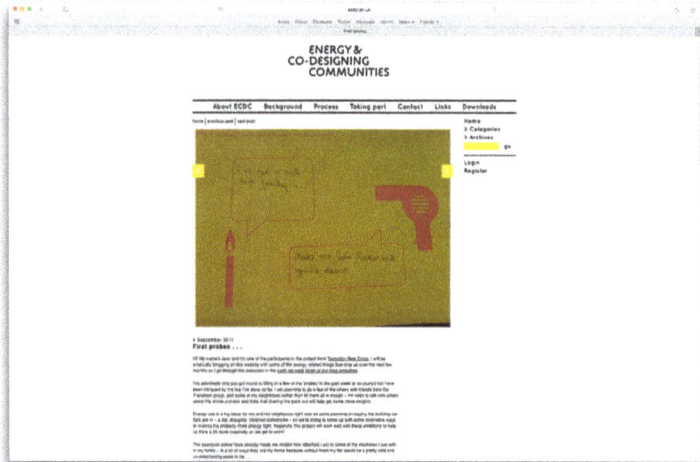

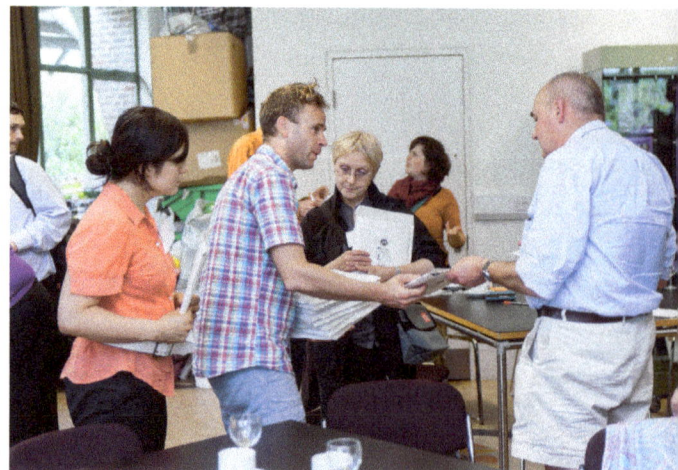

climate change, and also of their ability to laugh and play. As they left, we presented them

Household attitudes towards the use of energy were captured as a set of rules, which were in one case fixed to a fridge door

ENERGY RULES

1. Take on producers and organizations. Don't just focus on consumers.

2. Down with single-glazing.

3. IT'S STRICTLY Dancing

4. When in Rome, don't do as the Romans do

5. ALWAYS fart

6. Wait with washing as long as you can.

7. Throw out the -TV

8. public transport ONLY

9. NEVER eat imported food (or meat)

10. EVERY monday, don't shower

with Cultural Probe packs containing collections of new tasks for them to complete at

conversation between the two objects.

They tell me I should slow down or risk burn-out

You need to recharge my friend – get connected!

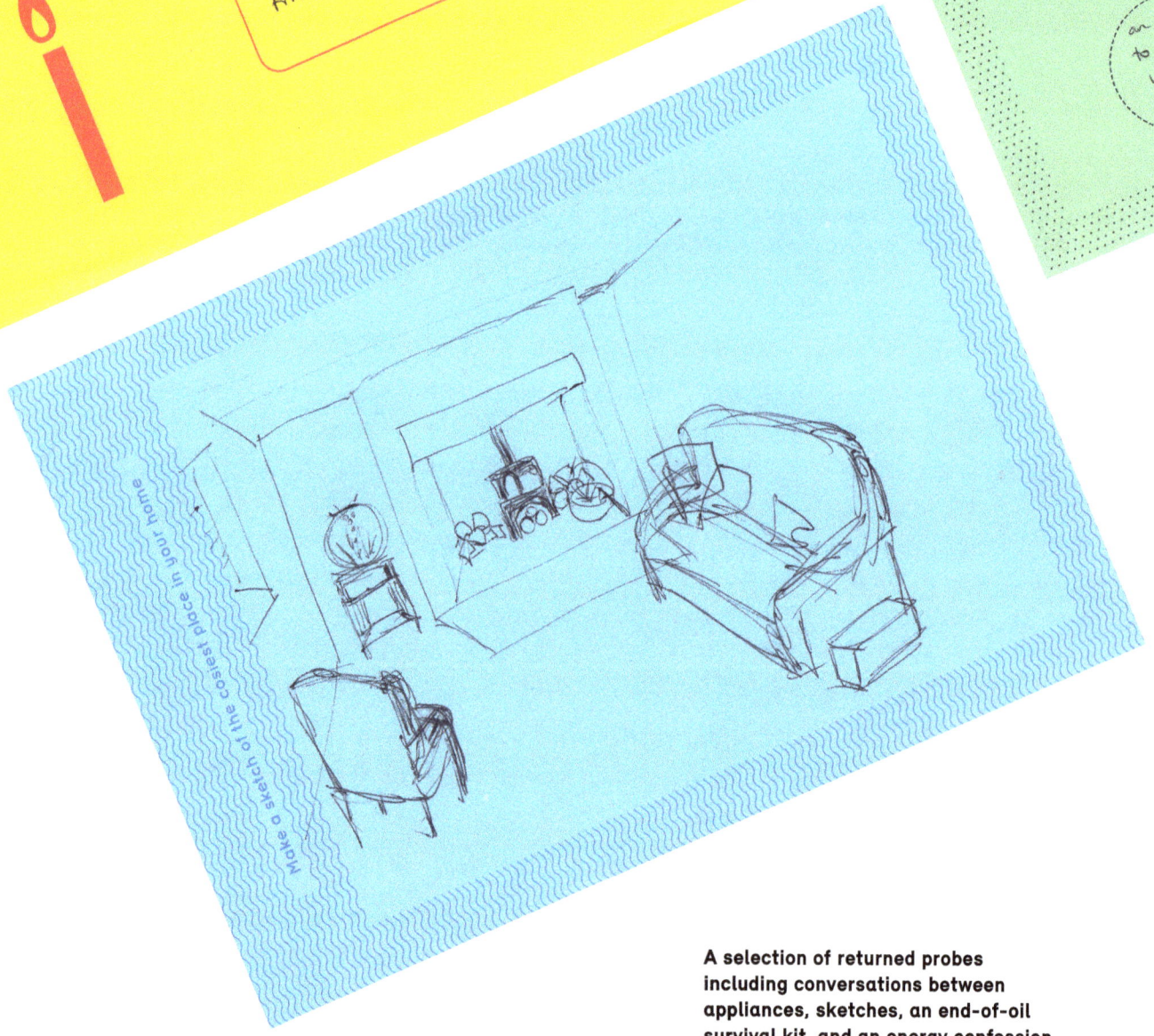

Design a survival kit for the future

This is for a

infinitely usable water purifying bottle

a rifle to kill rabbits etc & protect

an axe to chop wood.

Make a sketch of the cosiest place in your home

A selection of returned probes including conversations between appliances, sketches, an end-of-oil survival kit, and an energy confession

home. Again, the tasks were playful and perhaps surprising. What are your rules about

sitting next t...
...sic

...through

...ndow

future

ground source heat pump for heat

wind up torches/lamps for light

seeds
3 chickens
2 pigs

— steam

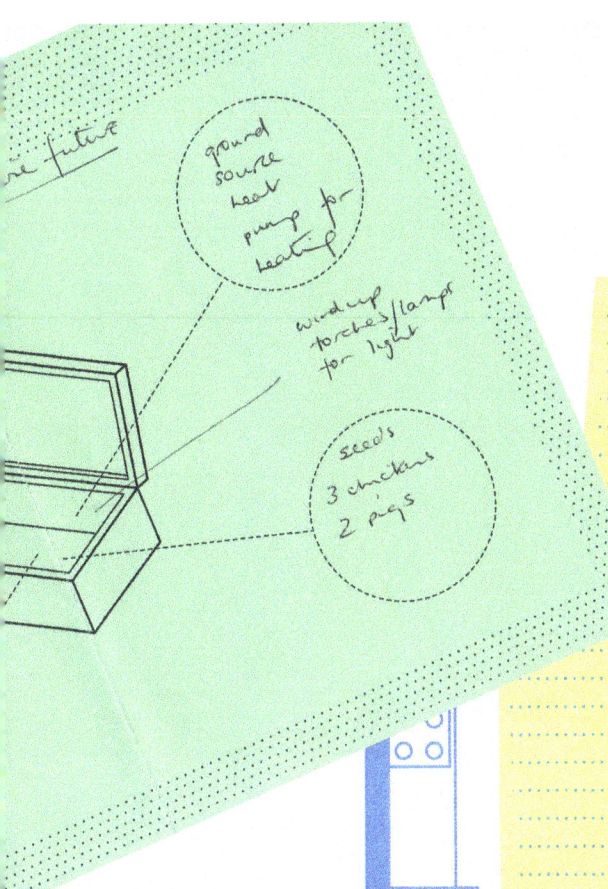

confess a guilty energy usage secret

take a hot bath every night.
put too much water in the kettle every time.
I don't know how much the lightbulbs use.
I don't own a microwave.
I sometimes take a taxi instead of a bus.
I fly almost every month. Actually. August
to november this year, I have flown
every month

Write an obituary for your favourite appliance.

write an obituary headline.

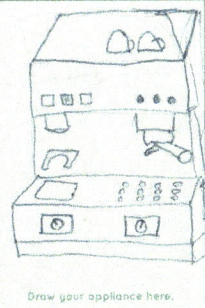

Draw your appliance here.

when we first met, you were abandoned and dishevelled. It was love at first sight. I wasn't sure what would become of us but I felt you were special. And you were special. I have never met someone with such a strong personality. Sometimes you would produce the best coffee in the world, nutty, smooth, thick and sweet. Sometimes you would throw coffee-dust all around the kitchen or splutter sour coffee everywhere except the cup. Then I tried to be kind to you. I thought you would live forever, but not even you, my Rancilio did. One day you went. Bye Bye, I will never drink coffee again.

energy? What would two appliances say to each other? As we waited for the probe returns,

My energy monitor: Chronicle of a failed attempt

Liliana Ovalle

Last month, I moved out of my rented flat. After living three years in the attic of a converted pub, I packed up all my belongings but one: an energy monitor. For the last two years, the device had been trapped inside the meter cabinet located on the ground floor of the building. It remains there, clamped to what I believe is the electricity meter of flat no. 7, my ex-flat.

Energy monitors are a key measure within current carbon-emission reduction policy. To better understand how these devices affect everyday life, we distributed a variety of monitors amongst the members of the ECDC team for a short trial. Covering a range of brands and styles, each monitor offered particular features, from a discrete minimal presence to a range of online services that could expand the device into a complex network of synchronised power switches and appliances, a gateway to the Internet of Things. With multiple graphs and quantifications accompanied by pictograms of smiley and sad houses, the monitors were presented as the ultimate tool to becoming an informed energy user. I received mine with excitement, eager to discover and quantify the flow of electricity going through my house, which until then had only been evidenced through quarterly bills.

As soon as I got home, I opened the box of my 'Alert Me Starter Kit'. The kit consisted of different electronic components: a battery-powered wireless transmitter that clips to the electricity meter, and a receiver unit that connects to the broadband router. After carefully reading the instructions, I began set-up.

My first task was to access the electricity meter. I had seen a locked door on the ground floor with signs indicating that it contained the meters and other installations of the building. I had been dissuaded from carrying out further explorations by a red sign which read 'Danger: Electricity shock risk'. On my first attempt to break into that semi-restricted zone, I found my first obstacle: the door was locked and I had no key. I had never before needed to open this door since electricity readings had occurred anonymously by means of an unnoticed visit by a reader sent by the electricity company. After looking for the key amongst all the appliances

Fire door
keep
locked
shut

manuals, loose keys, and other bits that the landlord had left us, I finally found it.

I opened the door only to discover a second obstacle: all the cables coming from the row of electricity meters were protected with plastic wiring duct. There was no way to access beyond this protection, and after failed attempts to remove the lid with a screwdriver I decided to squeeze my hand through the duct and try to clamp the monitor. This was a somewhat blind and uncomfortable manoeuvre as I wasn't sure which of the cables that I could feel corresponded to my flat – never mind that the electric shock warning kept flashing in my mind. It took me a few minutes of groping inside while I hoped that none of my

neighbours would appear and find me with my hand trapped inside the guarded electricity installation. I finally clipped the clamps to what I believed was the correct cable.

Excited about achieving what turned out to be the trickiest step of installing the monitor, I went back to my flat to check if the hub was receiving any signal, only to be disappointed to find there was none. After long chats with the supplier's online support, I was advised to remove the clamp to check if the signal of the transmitter was strong enough to reach my flat at the top of the building, which would require breaking into the electricity installation again. In the next two years, I never returned to it. The space behind the

locked door remained a restricted area to me, and I preferred to avoid the discussion with the landlord to get proper authorisation to access the space.

As I was packing to move out last month I had the uncomfortable reminder of the monitor inside the meter cabinet when I found the box with the rest of the components. I decided to leave the device behind and move on. I still picture it transmitting undetected signals to the world, trapped in the dark duct where it will probably stay until another tenant attempts to break in.

At least, we tried: one of the things we learned is how recalcitrant physical infra-

Rescripting monitors

This workshop encouraged researchers to challenge the behaviour of smart monitors, in order to develop fresh perspectives regarding the design of a technology platform for the communities.

use by non-technical or inexperienced users of monitoring technology discouraged or excluded

technology mediated access

renders user as completely dependant on online access

no physical presence

no reader or display

'cost of a cuppa' **individual** cost reduction as priority

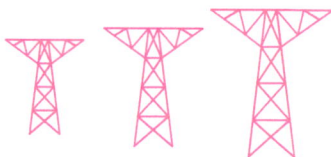

£

15g CO_2

the home/domestic as a **closed system**

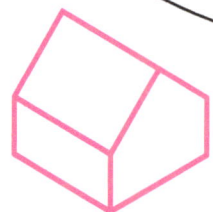

no relation to the national grid

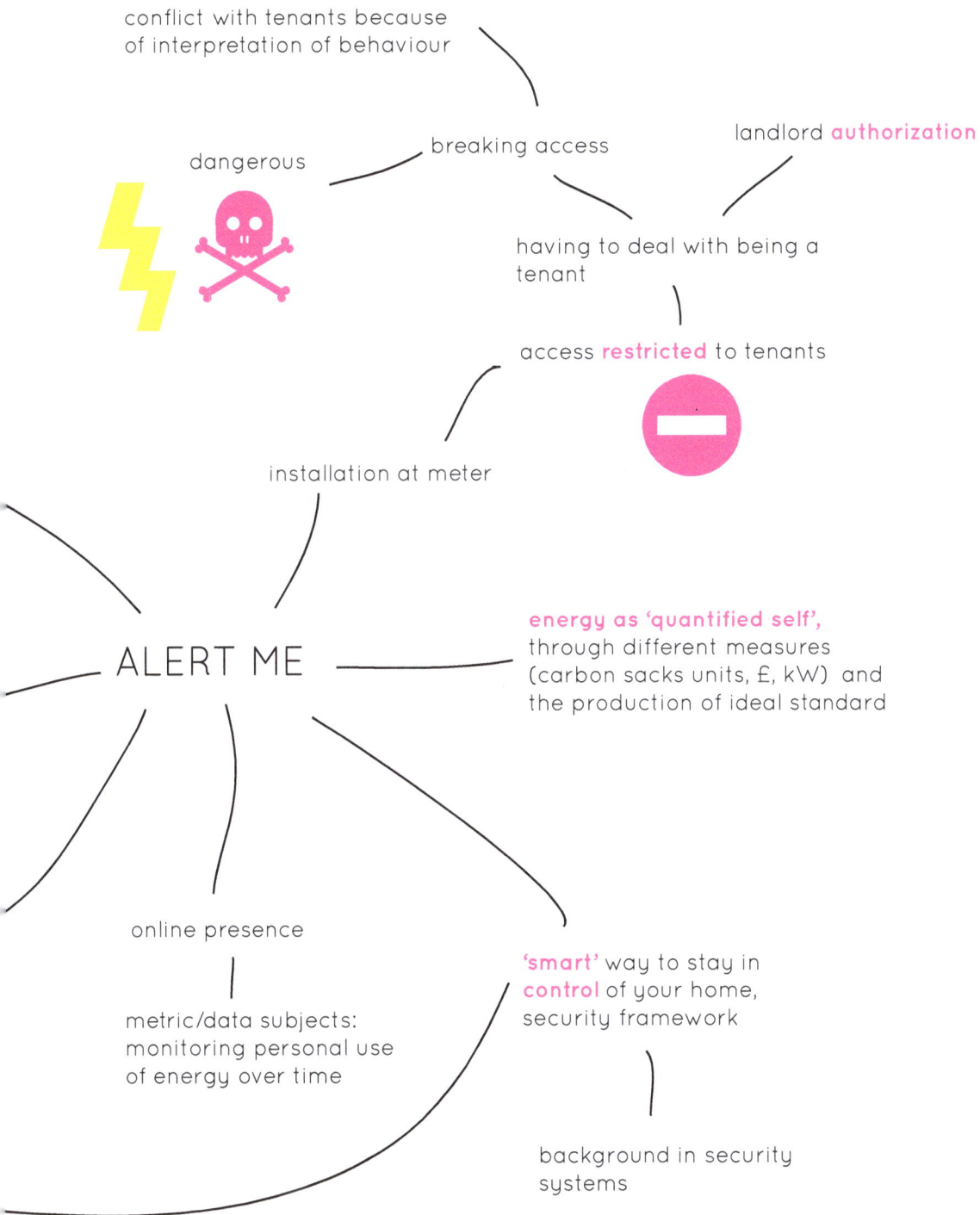

conflict with tenants because
of interpretation of behaviour

breaking access

landlord **authorization**

dangerous

having to deal with being a
tenant

access **restricted** to tenants

installation at meter

ALERT ME

energy as 'quantified self',
through different measures
(carbon sacks units, £, kW) and
the production of ideal standard

online presence

metric/data subjects:
monitoring personal use
of energy over time

'smart' way to stay in
control of your home,
security framework

background in security
systems

needed to 'rescript' them, by examining how the assumptions behind them portray their users and the issues, problems, and possibilities of energy-demand reduction. Finally, after

DESIGNING

As we became increasingly familiar with the energy communities, we started to imagine what we might make for them – something we had left almost entirely open when we started the project. Sketches led to conversations that led to more sketches and eventually to collections of concept proposals. As our ideas flowed, a tacit consensus formed that we would not be designing to reduce energy per se, but rather to address the communities and their situations. The sketch proposal for an 'energy babble' seemed to capture the trend of our ideas and, along with its translation into a textual design brief, set the course for further development.

Over the next months, we worked on the software, electronics, sound, and product design of the system in parallel. This was not a matter of finding solutions to a technical specification. Instead it was more like sculpting in these media, working them to find the final form of the Data Catcher. The project took surprising twists along the way – impassioned conversations about coiled cables, deep thinking about simple musical jingles, the commissioning of laboratory glass blowers – until the final Energy Babbles, unexpected and idiosyncratic, became real objects sitting in our studio.

Workbooks

After the initial engagement we started to explore ideas for the systems or artefacts. These ideas and insights were captured in workbooks, which collated sketch proposals, insights, and articles, that together create a framework for the project. From energy tourism, and insistent activism to energy awareness, the workbooks high-lighted areas of interest that helped us to identify potential directions and themes, creating a design space that led to the creation of the Babble design brief.

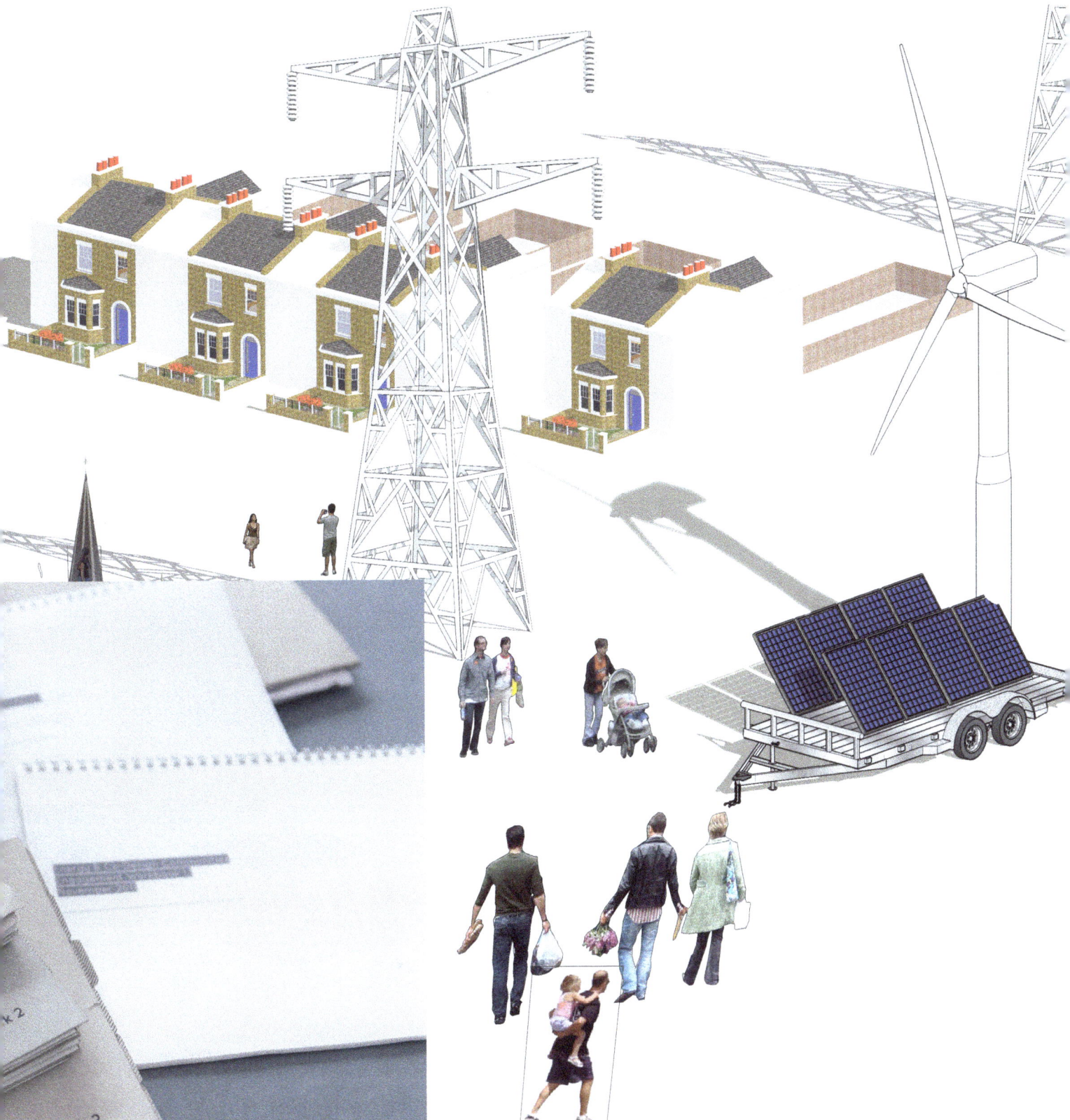

Sketch from a workbook proposal, showing potential interventions in a community pub

We produced workbooks that collected tens or hundreds of design proposals, each an evocative

Workbooks capture insights and ideas that emerged from the initial engagement, often leading to evocative proposals

combination of images and text that pointed to a direction for design. Thematically

UNLEASHING: RITUALS OF INITIATION

"Some groups, such as [Transition Penerth](http://www.transitionpenerth.com/?TP"), started pretty much from cold with an Unleashing, because they had the opportunity of having Richard Heinberg present it, a rare enough possibility. The ideal though, as I see it, is a bit like one of those toy volcanoes that children like you gradually add a bit of vinegar, a bit of baking powder, a bit more vinegar, a bit more baking powder, until the pressure inside builds to an unbearable point, and then BAM, there is the Unleashing. It marks the arrival of the project, and it is a celebration of the community's desire to act."

http://transitionculture.org/2007/01/22/first-steps-for-a-transition-town-initiative-3-the-official-unleashing/

Image above: Kids Lab Volcano Making Kit

ENDLESS BOREHOLE

Invisible technologies invite imaginative work. We stood at a location in the school playing field where the ground source heat pumps was installed. But how deep is the pump buried? Opinions varied, 60 meters, 600 feet, 600 meters?

Answer: The borehole collector for the Vaillant System is sunk 100 meters deep.

Crust
Moho
Upper mantle
Lower mantle
D" - layer
Outer core
Liquid-solid
boundary
central heating

Above right: adapted from Guy Keulemans, "Dumb Probes & Nuclear Fuel, Sinking to the Centre of the Earth, Melting Rock and Iron"

TECHNOLOGY SAVES LABOUR

Image above: Cosi E wearable electric blanket

PRECIOUS METALS

COINS

CATTLE

COWRIES

REMOTE VALUES

Like the Rai Stone sunk under the sea, certain values related to energy remain intangible or remote.

eco gadget library

Borrowing and sharing devices

above: West Oxford Renewables on Ecomode and Hastings Library

Image above: Cosi E wearable electric blanket

UNEXPECTED SWAP

Tuesday 10.22 am   Sunday 10.34 am

CLOTH CUPBOARD

The renewable displays are always in the back of cupboards, protected by folded linen and towels. Approaching the technology is to trespass upon the clean, flower scented whites.

CONVEX DOMESTIC

"I went a bit extravagant with the shower"

Above left: At home http:// archiveni.org/furniture/furniture/what-a-death-type-are-you/

REPORTING

UNPLEASANT TO THE EYE TODAY IS THE FUTURE OF WORLD HERITAGE

COMMUNITY IS A WORD I TOO MUCH

JUST ACCESS TO THE NEEDY THROUGH THE MINISTER OF THE COMMUNITY, A POSITION TAKEN BY GOD ALL THE TIME

PUBLIC ENERGY POOLS

Somewhere in New Cross   Somewhere in Ladock

Public, situated displays that are tied to the same pool of energy. Using the display in New Cross pulls energy from the shared resource and reduces the amount of energy available to others.

ENERGY GUILT-CLOUD

ENERGY GUILT-CLOUD

Unload your guilty energy secrets in private

Image above: Happylife, James Auger and Jimmy Loizeau
http://www.auger-loizeau.com/index.php?id=23

TOURIST POWER

ENERGY GUILT-CLOUD

That's not my name

Bill Gaver

They call me 'Bell'
They call me 'Stacey'
They call me 'her'
They call me 'Jane'
That's not my name
That's not my name
That's not my name
That's not my name...
The Ting Tings

One of the recurring challenges of interdisciplinary work is in handling the interpretations of one's work that come from those outside one's field. This is probably true for any discipline (my wife continually needs to reassure people that she's 'not that kind of psychologist'), but seems particularly acute for design. This may be, in part, because design's public face in shops and magazines gives the impression of being easily readable, and the basic activity of creation and development seems so universally human that saying we're all designers is a well-worn trope (though to paraphrase Bill Buxton, if we're all designers because we choose what clothes to wear, we're all engineers because we can change

a lightbulb). In any case, the hugely variegated practices that often lie beyond design's immediately accessible face – differentiated in motivations, conceptual underpinnings, processes, values, expertise, audiences, outputs, and so on – seem to blur together from a disciplinary distance, so that it is not uncommon for our practice in the Interaction Research Studio (IRS) to be attributed with features we don't recognise, or lumped together with practices to which we don't relate.

So it was that we found ourselves, early in a very large project (not this one), working with a variety of computer scientists, sociologists, psychologists, and human computer interaction specialists, having to explain that we did not just want to make their prototypes pretty. Instead, we explained, we could play a first-class role in the conceptualisation and implementation of new devices and systems, and moreover we might bring a distinctive approach to this that would complement the work of other disciplines. Later in the same project, we introduced 'cultural probes' – collections of evocative tasks distributed to elicit informative and inspiring responses – only to be told they amounted to nothing more than 'ethnography in a dress'. Apart from the implicit sexism – we explained – this label overlooked the very different epistemological commitments of the probes, which balance the grounding offered by empirical encounters with the mutual confusion and interpretation created by open-ended and even absurd tasks.

Other claims are more irksome to counter. For instance, we have been credited with (accused of?) making art, not design, on the grounds that we: a) do not have clearly identifiable clients for our work; b) do not practise in a commercial context; c) do not seek to solve problems or address identifiable needs; and d) sometimes base our methods on artistic practices rather than those of social science. Although we can address each of these attempts to define borders around design, it is more difficult to define clearly the distinction between art and design, because of art's extraordinary ability to annexe ways of working, or forms of output that have been suggested as quintessentially non-artistic. The most satisfactory response, we have concluded, is to point out that just as Duchamp's Fountain is a urinal made art by its setting in a gallery, our work is design by virtue of its intention and circulation in communities of practice associated with design, but this still doesn't always appease critics who base their judgements of art/non-art on appearances.

More problematic still is the identification of our design work with speculative or critical design (SCD), which has occurred frequently (and even in this book). Like many design genres, these approaches are defined largely by examples, practitioners, and somewhat sloganistic definitions. For instance, SCD explicitly sets itself in opposition to 'affirmative design', which 'reinforces the status quo'. Critical design tends to work with the potential of current trends, creating 'design fictions' that explore implausible (and often dystopian) values for existing technologies; speculative design, in contrast, tends to extrapolate possible futures from present realities, creating fictional scenarios in those futures and finally populating those scenarios with designed artefacts that reflect their implications. In both cases, design is considered a 'tool to create not only things but ideas', and thus it is not necessary that an artefact actually function technically or be encountered by its putative users. Instead, critical and speculative designs are valued for their ability to be communicated in striking ways (e.g. in galleries or the press), and to provoke reflection and discussion about the assumptions they address. This is often achieved by creating a form of controlled ambivalence, in which the appeal of

a well-crafted device temporarily masks more ominous and disturbing implications.

Given that IRS work also tends to counter assumptions prevalent in technology design, to privilege unusual values and activities, and to embrace playfulness and what James Auger refers to as 'irreverence', it is perhaps not surprising that it is often identified as a form of SCD. This obscures fundamental differences between SCD and our practice, however, that we believe are crucial to understanding the Studio's work. For instance, technical function and lived encounters are of low priority for SCD. This is symptomatic of SCD's agenda to, primarily, critique or at least interrogate current assumptions and their potential impacts in designs intended to be thought-provoking and even disturbing. IRS designs, in contrast, are usually intended to be usable by and engaging for their intended audiences, without any backstory or unpacking of the assumptions they address. Thus, integral to our work is the production of working research devices, and their deployment for long-term field trials involving extended participation periods. These trials anchor speculation to empirical encounters, allowing us to assess how people actually engage with our designs rather than leaving this to the imagination. Often, the role of participants in co-creating the meaning of our designs is enhanced by creating designs that are open-ended (rather than ambivalent), and capable of supporting many possible interpretations and engagements in ways that can be as revealing of their users as of the designs. Finally, through this commitment to fully finished, functioning devices that participants live with over time, the IRS disavows any supposed opposition with 'affirmative design', and instead seeks to expand the repertory of technology design to embrace new values, activities, and techniques.

Of course, one tactic for countering the all-too-frequent identification of IRS designs as speculative or even critical would be to subsume our design work under a distinctive 'brand name' of its own. (Situated Design? Open Design?) That we haven't done this – apart from occasional, and now largely historical, references to 'Ludic Design' – is not just a reflection of our poor imaginations, however. (Exploratory? Unsettled?) Instead, it seems impossible to find a term that would capture a practice which, while arguably having a distinctive style, has ranged considerably. Such a term would most likely be too generic to communicate well, while conversely placing potentially unnecessary constraints on future evolution. Moreover, we are reluctant to join the competitive market of branded design approaches, vying for publicity and followers. Finally, drawing a boundary around our design work to distinguish it from others (including SCD) would also have the undesirable consequence of separating it from the 'normal' design with which we would like to interact, and which we would like to affect. Thus, we will continue to avoid branding our design approach, letting our methods and the things that we make speak for themselves.

and unusual, debates ensued (and continue in this book) about whether they should be

ECO-BABBLE

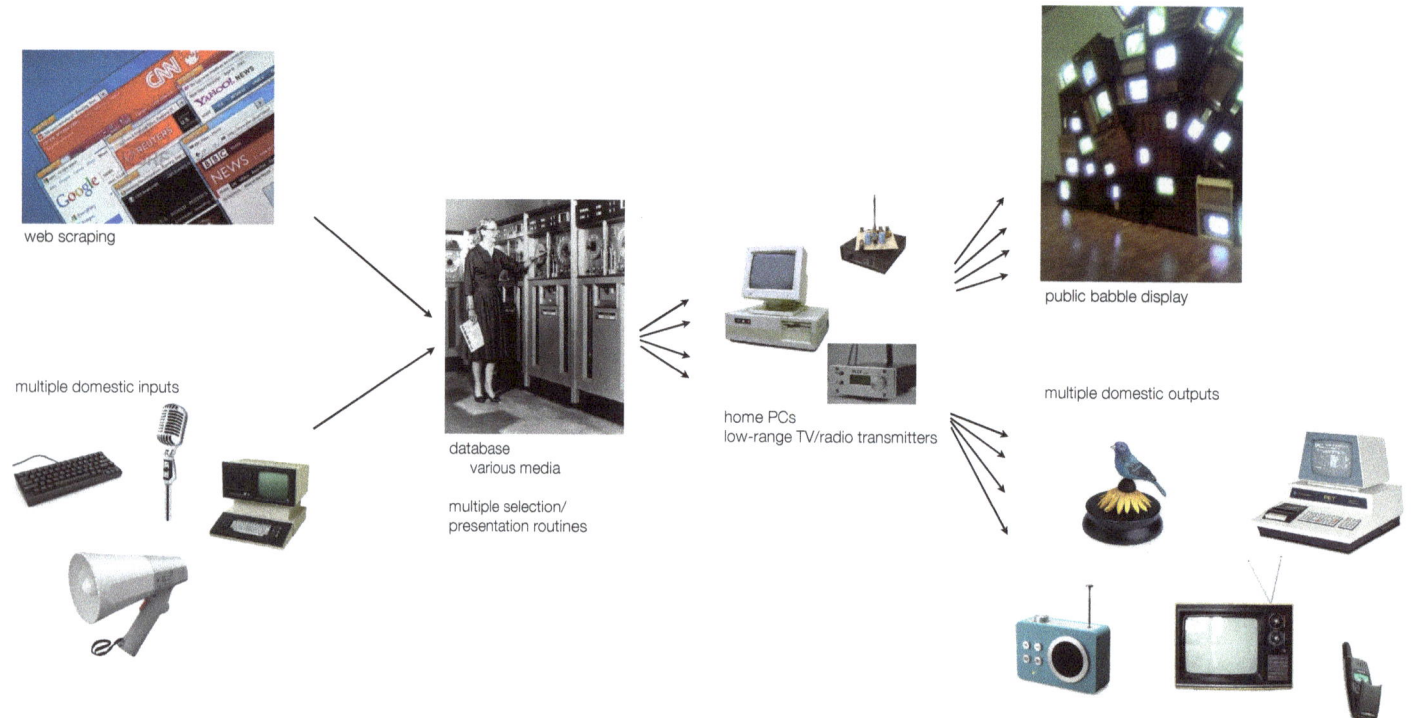

web scraping

multiple domestic inputs

database
various media

multiple selection/
presentation routines

home PCs
low-range TV/radio transmitters

public babble display

multiple domestic outputs

Proposal & Brief

The original Eco Babble proposal: A system that would gather data from a combination of web scraping and participant input to a central server. From there, it would be organised and processed for redistribution by a variety of public and domestic devices spread across the energy communities.

This was re-expressed as a brief that accrued amendments and alterations as it circulated amongst the team. Though it became messier in the process (we have actually cleaned up out-of-order and duplicate footnote numbering here) this didn't matter while it served as a forum for a living conversation rather than a resolved or archival artefact.

considered 'speculative design'. Finally, a key proposal resulted for an Energy Babble, and

Energy-Babble

Design an Energy[1]-Babble[2] system[3] that displays[4] material[5], collected[6] from some combination[7] of individual[8], community[9] and public[10] sources, to open[11] and promote constructive[12] affect[13] and involvement[14] in energy reduction[15] issues and orientations[16]. More specifically, the system should support understandings[17] of, and practices related to, energy demand reduction.

---

[1] Don't forget that energy reduction discourses might include those that are not part of the 'green' agenda, e.g. climate change deniers, travel writing, critiques of wind generation, advertisements for luxury goods, sports car adverts.

[2] The intention is to contrast inconsistent or contradictory ways of talking about the environment (within as well as, and more importantly than, between pro- and anti-green' agendas) to provoke uncertainty and reflection. Isn't it?

    The intention is to draw on discourse as a means of mediating the environment. This includes different and contrasting ways of talking about the environment, not only to provoke uncertainty and reflection but also to render the actors and issues implicated in environmental 'babble' open to investigation, scrutiny, and reflection. The intention is also to go beyond mere discourse. Here, 'babble' also refers to a way of bringing into being new environmental actors through mistranslation and mash-up.

[3] I am assuming this will include[18] distinct front[19]- and back[20]-end considerations, which may be considered together or separately.

[4] These could include domestic devices, mobile ones, ones designed for specific environments (the shower? the car?), as well as public displays, themselves potentially designed for specific places (public hall? library? notice board? park?).

    A system that will support different modes and sites for interaction between communities and environmental babble.

[5] Material might include text, images, sounds, videos, or god knows what.

[6] Through people's purposeful input into the system, or web scraping, or bots, or...

[7] Single source systems might also work, but seem less likely to create a Babble[2].

[8] For me, of particular interest is the idea that users' responses to things they encounter on the system themselves become material for the Babble.

[9] For instance by scraping websites, mission statements, blogs[21] and taking some forms of community input[22] into the system.

[10] Policy docs, news reports, blog posts etc - but perhaps also by opening the system to input from individuals outside our core communities?

    A more expanded notion of the public, or publics. One that includes different human environmental actors, such as community members, practitioners, policy representatives as well as non-human actors – such as bots etc.

[11] My understanding is we're trying to unsettle people's understandings of the environment, potential problems concerning it and the right ways to address those problems. We may — or may not — benefit from considering what those understandings are, and the angles from which they might be challenged.

    The intention is not only to unsettle people's understandings of the environment, it is also to... (something to do with the type of things or agencies that we recognize as being part of environmental issues.

[12] By constructive we mean to refer to a 'constructivist' process where environmental objects and practices are reported on, created and given lives of their own within the system. This is in contract to a 'critical' perspective.

[13] Or not. Perhaps we just want our designs to act as 'the idiot', insisting that there is something else that matters, in a way that doesn't readily lead to coherent reflection.

[14] Involvement would also entail communicational processes with other users of the prototype: involvement can therefore connote the contingent making of community/ies.

[15] This footnote has intentionally been left blank.

[16] Not just the way people talk but also the ways they act.

[17] Presumably we would want multiple understandings but also practices leading to understanding through collective interactions? Also understanding feels a bit cognitive when perhaps we want a more general sense that opens up possibilities of understandings?

[18] There is also much potential in stuff in the middle, e.g. automatic prompts for input (the Eliza system ideas) and translators and filters for output.

[19] Keep in mind production practicalities. We might want different physical devices housing the same basic infrastructure. Equally we might want identical devices with channels, or settings, or configurations that distinguish them.

[20] A database driven system? Some sort of web interface?

    Can the database itself become some form of environmental object?

[21] Including those travelling between communities, i.e. the Songbird proposal.

[22] Are there other ways to get 'community' inputs? Are there ways to reflect both the fact that we are working with self-identified 'communities' and Matthew's point, that communities are dynamic and emergent?

Why an 'Energy Babble?'
Bill Gaver

The Energy Babble first appeared as a proposal labelled 'eco-babble' that was presented as a collage/diagram on a single sheet of A4.

When discussed at one of our design meetings, it quickly became accepted that this was what we would make. This is an experience we have had in numerous other projects, wherein what seems like a vague and confusing space of possibilities collapses into a more clearly focused sense of direction.

The Energy Babble idea 'worked', in part, because it brought together suggestions and explorations from other proposals. These ranged from the idea of whispering accounts of energy practices to the 'energy shrines' that would distribute them to other communities, to the idea of 'preparedness advice for the end of oil blaring out' of a device to be mounted near energy meters, to poetic treatments of nomad devices that would share stories. The Energy Babble consolidated these lines of thought in a proposal that suggested how they might be combined and achieved practically. In effect, it served as a hinge between our design explorations and more focused development work, providing initial guidance for this next phase of work.

But why were we drawn to the idea of an Energy Babble?

In part, it was because it built on expertise we had been developing over a number of projects that involved reframing content drawn from the internet – a strategy that we knew could yield rich results. Building an Energy Babble would mean dramatically expanding the number of sources we would draw on, as well as developing ways to communicate to a central server from devices – both challenges that would expand the Studio's capabilities.

We also liked the way it combined ideas about letting communities report to each other about their practises with notions of providing new information about energy policy and technologies to the communities.

But perhaps most of all, the concept of an Energy Babble seemed to reflect the situations in which the energy communities found themselves. Getting to know them, we had become aware of the difficulties they had in dealing with complex and changing governmental policies, with rapidly developing technologies for saving and producing energy, with engaging members of their wider communities, and with communicating amongst themselves. Further, it was evident to us that they did not share the same understandings and assumptions about what they were doing. Some of the communities were concerned with energy reduction to prevent environmental catastrophe. Others wanted to achieve post-oil energy self-sufficiency. Others wanted to generate energy as a source of income. Each had its own understanding of motivations, issues and approaches, and this meant that, while it seemed from afar that they were working in congruent ways, in reality, they were all talking about slightly different things.

The Energy Babble seemed a satisfying response to the communities' circumstances. It wasn't really conceived as a source of new information to the communities or as a communication medium for them to share (though this later became the way it was often presented within those communities). Instead, it appealed as a kind of playful mirror to hold up to the communities, one that would reflect their complex and confusing situations, perhaps even in a humorous way.

That the Energy Babble was not conceived as a traditionally utilitarian tool did not mean that it was a critical or speculative design, however. While it embodies some rueful headshaking about the seeming impossibility of the communities' pursuits, it wasn't intended to mock them or to paint an overly critical or bleak picture. Instead, we meant for it to engage them in a kind of in-joke about the absurdities they faced, and ideally to prompt them to think about alternatives. We also anticipated that a Babble could be engaging and pleasurable in its own terms. From this point of view, the Energy Babble was intended to be as instrumental and functional as any traditional design, however untraditional its purposes may be.

Design and development

There were four dimensions to the design and development of the
device: hardware, sound, software, and form. We worked on each in
parallel, with frequent meetings to ensure integrated progress.
Numerous experiments with early hardware prototypes, Twitter 'bots',
musical phrases borrowed from various sources, and a stream of wood
and plastic form studies helped us along the way. We knew we would
batch-produce the resulting design, so all this activity became a means
to transform the initial sketch of the system into a set including, amongst
other things, carefully detailed computer code and specifications for
glass blowers and injection moulding specialists.

Hardware

The underlying hardware platform was informed by our design brief. We considered that the client devices would support audio recording and playback, and connect to the internet over Wi-Fi for sending and receiving audio files. The device was developed incrementally, and used mainly off-the-shelf components; though we did design some bespoke circuit boards to support the physical interface. There was a variety of small computers that could have supported our requirements, though we settled on the Raspberry Pi due to its low cost and the support of a large and enthusiastic community of developers.

A drawing of the final hardware system depicting the main elements and their connections to the Raspberry Pi and USB.

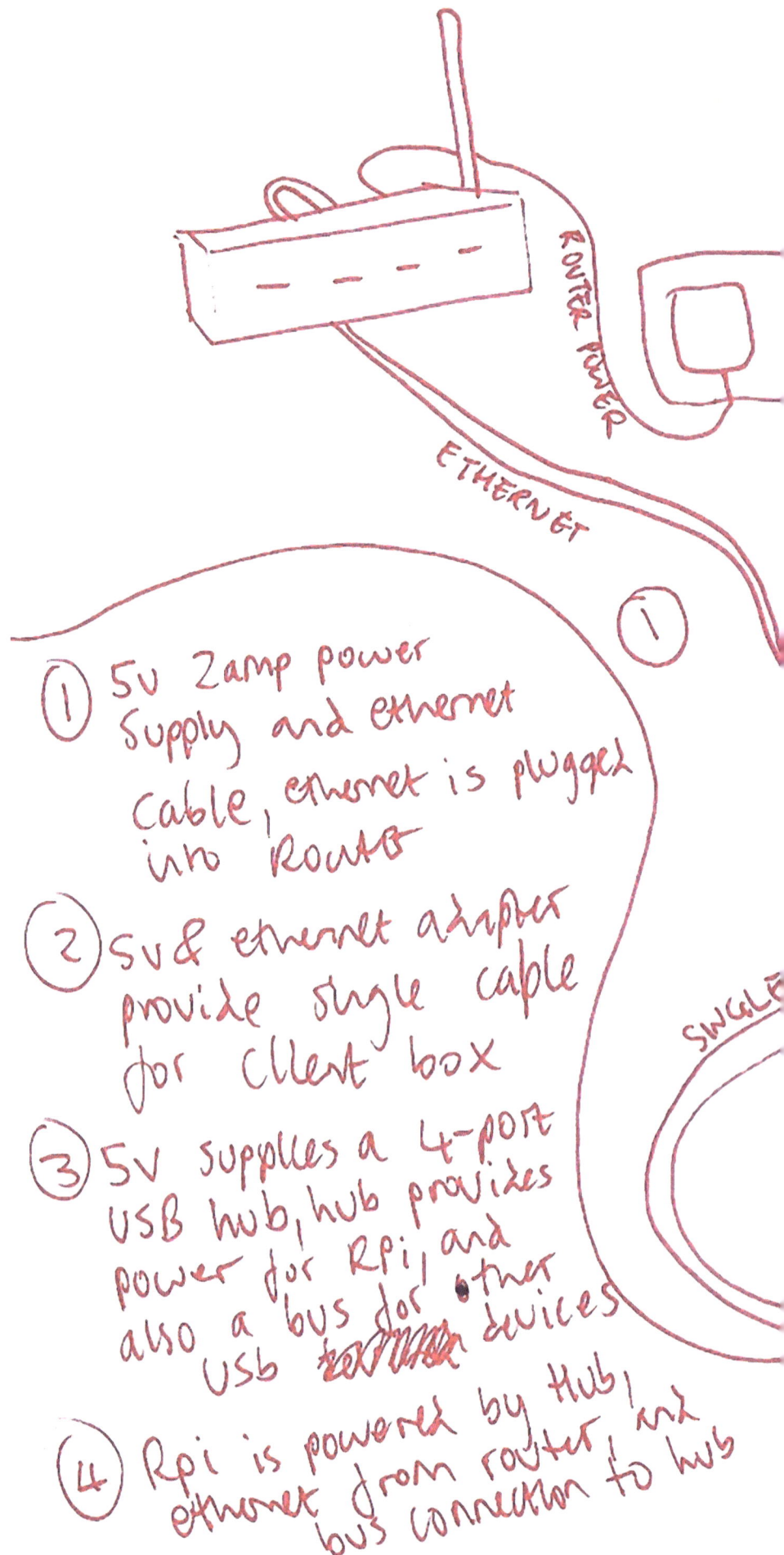

ROUTER POWER

ETHERNET

SINGLE

① 5v 2amp power supply and ethernet cable, ethernet is plugged into Router

② 5v & ethernet adapter provide single cable for Client box

③ 5v supplies a 4-port USB hub, hub provides power for Rpi, and also a bus for other USB devices

④ Rpi is powered by hub, and ethernet from router via bus connection to hub

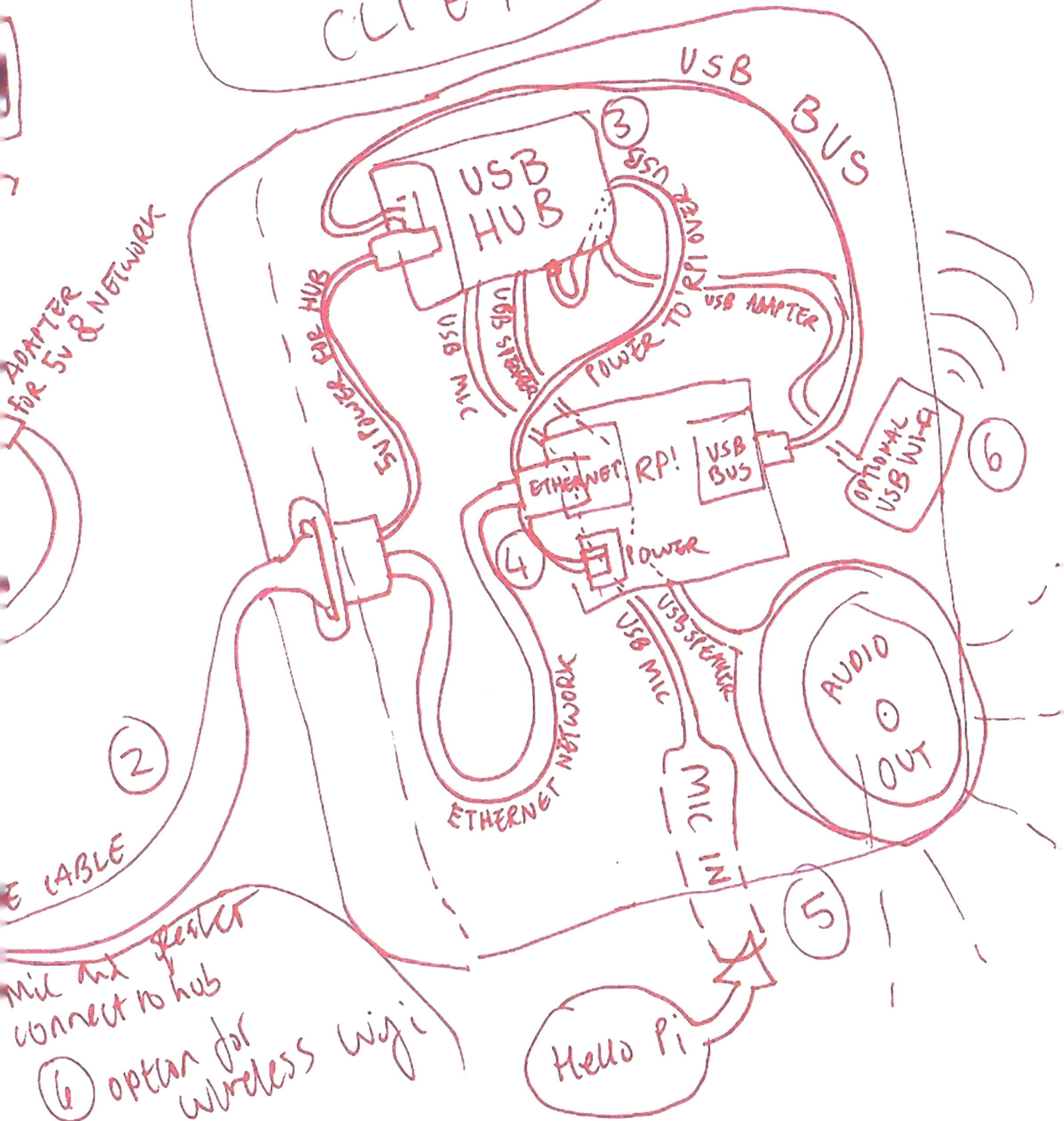

BABBLE CLIENT

USB BUS

USB HUB

3

USB HUB

USB SPEAKER

USB MIC

USB SPEAKER

POWER TO RPI USB ADAPTER

ADAPTER for 5v & NETWORK

ETHERNET RP! USB BUS

POWER

OPTIONAL USB HUB

6

ETHERNET NETWORK

USB MIC

USB SPEAKER

MIC IN

AUDIO OUT

2

E CABLE

mic and speaker connect to hub

6 option for wireless wigi

Hello Pi

5

the Raspberry Pi micro-processor platform that would allow us to push information between

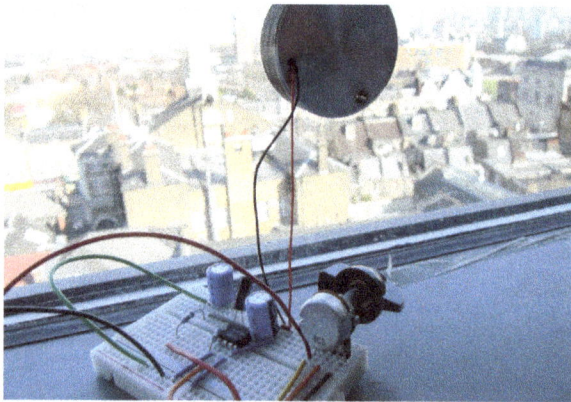

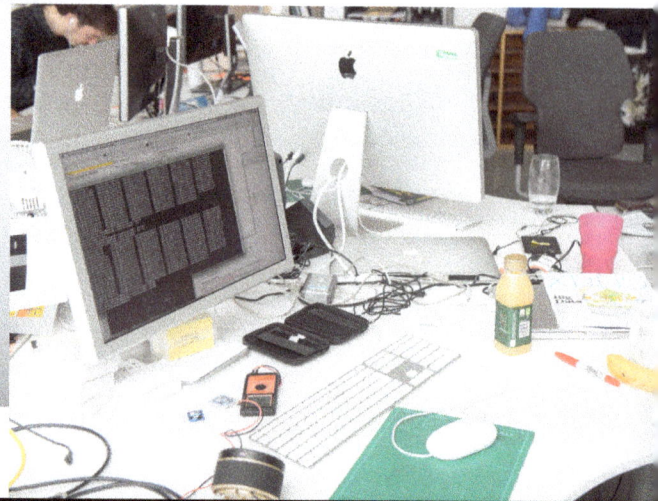

devices and a server, the synthesised voices that would report the news, the speakers

CotEditor   File   Edit   View   Format   Text   Find   Window   💲   Help

.F ⌄ | Unicode (UTF-8) ⌄ | Shell Script ⌄ | | Designing 55
dings | File Encoding | Syntax Style

This script uses differe...voices to tell guilty secrets ⌄ | ⌄

```sh
#!/bin/sh


# This script uses different voices to tell guilty secrets


$guilt1 = "Sometimes I like to dry my body with a hairdryer."


$guilt2 = "Some times I leave the lights on in my studio just because it keeps my


$guilt3 = "I still leave the tap running when I brush my teeth. That's what's bother


$guilt4 = "I have the heating on when its hot outside."


say -v Serena $guilt1
say -v Moira $guilt2
say -v Daniel $guilt3
say -v Lee $guilt4
```

Serena — female, British English voice
Moira — female, Irish English voice
Daniel — male, British English voice
Lee — male, Australian English voice

Sound

Sound is an essential part of the Babble system. The devices would be embedded into domestic and public environments and would manifest like a kind of auto-mated talk radio station. It was therefore critical for us to attend to the design of the soundscape with as much care as we had the computational infrastructure or the physical casing. Synthetic voices were tested, tweaked, processed, and harmonised. Daily 'programmes' were scheduled and choreographed. News 'headlines' were read out at the top of the hour, and jingles were developed to provide continuity and rhythm throughout the day.

Early experiments with text-to-speech code to create anonymous voices for energy guilt confessions supplied by the communities.

who would let the device have a voice in the first place. We also started composing short

Sound design
Alex Wilkie

A Korg MicroKEY 37 USB MIDI keyboard was used as a tool to explore and develop the jingles of the Energy Babble soundscape.

Early on in our design process we recognised that the Energy Babble was going to be some kind of sound device that sourced spoken content from the internet. Not only would this inform and shape the design of the artefact's physical and visual form, it also meant that one of the key design challenges would be to compose the sonic characteristics of the device. This gave rise to a variety of design problems. How, exactly, should textual content drawn from the internet sound? What role can sound play as part of an interactive computational appliance? Could interaction with the Babble be focused around a sound interface? How can we sympathetically mix synthesised spoken word – generated from text-to-speech software – with music? Indeed, where do we, as designers, take our cues from in terms of music and sound design?

The studio has designed many interactive computational artefacts that are typically a mixture of industrial or product design as well as visual interface design – where various kinds of screen display textual and visual content – dot matrix displays, LCD displays, and so on. The Babble was going to be different. All of the content would be presented or inputted via sound. Tobie did some instrumental work in getting the text-to-speech software working on the Raspberry Pi-based system. He managed to put together a working system that included an automated online transcription service – which we were convinced was some kind of Mechanical Turk set-up rather than an advanced speech recognition service – as well as Apple™'s built-in synthetic voices to read out text. With this, and the emerging assumption within the project team that the device may well be left on all the time in homes and at work, for example, one source of inspiration we looked towards was radio. Here was a model of sound design that incorporated spoken content with sound, with the sounds often acting as cues for certain types of content or indicators of time

or other such information. Likewise, the design of radios as material objects also informed – or not – the physical design of the device, as Liliana Ovalle describes in her account of designing the device.

I volunteered to take on the job of designing the sonic 'personality' of the Babble. I was, admittedly, apprehensive about this, not least because of Bill's early renown for designing the SonicFinder as an auditory feature of the Mac OS interface in the late eighties. Given the high degree of finish that characterises much of the Studio's designed artefacts, I knew that not only did the Babble have to match these expectations in terms of design, but it also had to sound 'like' its physical form and be sympathetic to the timbre of the spoken content.

My first encounter with sound design came in the form of a musical umbrella that I designed for my graduation from the Computer Related Design Masters course at the Royal College of Art in 1999. Previous students on the course had done pioneering work in terms of tangible musical interfaces, notably Dominic Robson and Mark McCabe's 'Sound Toys', which set a precedent for designing experimental computer interfaces for music making. Back then, I was reading (or more likely misreading) *We Have Never Been Modern* by Bruno Latour (1993) and I was interested in taking a seemingly natural phenomenon – raindrops – and using this as a basis for a playful interaction between a 'user', a device (an umbrella), the natural environment, and sound. The 'Pedestrian Leisure Prototype' sensed the impact and pressure of raindrops, using piezoelectric sensors and the tensile capacities of the umbrella canopy to trigger MIDI (a standard protocol for allowing computers and instruments to communicate) sounds that were then modulated by sensor readings of the pH level of the water (by way of a sensor mounted at the top of the umbrella) and the

movement of the person carrying the umbrella – using a torus-shaped and water-based movement sensor in the handle. The sound produced by the umbrella was generated using Max/MSP, a graphical programming environment for audio, and resembled 'minimalist' music where harmonic patterns emerged through the patterning of individual notes. At my graduation show, the umbrella was exhibited as a working prototype using a foot-sensor-triggered shower unit to provide raindrops. After this, and whilst working as a designer, I formed part of a team of industrial designers, interaction designers, sound engineers, sound designers, and composers which was developing an interactive sound device to manage and improve the audio environment of office workplaces. Again, this project included much work with Max/MSP, which was the domain of sound designers and composers. My role, however, alongside a colleague, was to design and implement the graphical interface for the device – a shared user-interface that people could use to operate the device and control the sound through a web-based or Pocket PC (this was pre-iPhone) interface. The interesting challenge, here, was to design a cooperative interface that polled people's preferences rather than responding to direct input.

A number of constraints to the task of designing the sound for the Energy Babble quickly became apparent. First, the technical limitations of our Raspberry Pi-based system meant that we had to rely upon short samples rather than onboard generated sound so as not to take up too much memory, network bandwidth, or ask too much of the limited microprocessor. Second, by drawing on the vernacular of talk radio, the sounds would have to be short and expressive – in other words 'jingles'. Third, producing jingles invariably meant experimenting with and using MIDI sounds as the main musical resource, rather than the costly and time-consuming process of recording and sampling live musicians.

This became a mini-project of its own, as Alex explored different sources for inspiration

To go about designing and producing jingles, I first had to set up a working environment for exploring and producing the sounds. This included acquiring a relatively inexpensive USB MIDI keyboard (a Korg MicroKEY 37) as well as Logic Pro, an industry standard audio workstation and MIDI sequencing software application. Rather than the visually object-oriented paradigm, exemplified by Max/MSP, I would be using the other key paradigm for software music production to make the jingles – namely the visual sequencing of samples and MIDI instruments. Logic also gave me access to a whole raft of software-based instruments with which to compose the jingles – though I use the work 'compose' cautiously here as I in no way see myself as a composer.

I next had to think about where to begin. What kinds of talk-show radio should I listen to to gain inspiration for the kinds of jingles associated with different kinds of content. In conversation with Matthew, news jingles emerged as an obvious place to start. We scoured the web for examples, but we were mainly disappointed with what we found – simple, alerting, arresting, and officious-sounding alerts.

What did come to mind, for whatever reason, during this initial exploration of the jingle-sphere, were the movies of the French comedian and filmmaker Jacques Tati, particularly those that featured his celebrated character 'Monsieur Hulot'. In films such as *Mon Oncle* and *Play Time*, mass-produced technologies such as domestic appliances, as well as workplace settings emit estranging hums, rings, and buzzes that render the (for the time) new post-war technological environments of everyday life exceptional, uncanny, and yet playful. This combination – the ludic estrangement of routine technological artefacts and settings – seemed apt for informing the sound design as well as reflecting the material design of the Babble,

which had, at the time, progressed to a form close to its final specification.

Another inspiration, related to the everyday technological soundscapes depicted in Tati's films, came from public address (PA) systems that are commonplace in mass transit systems and the built environment, and particularly the way in which soundscapes are 'branded' with organisational and institutional auditory identities. Another French example this evoked for me was the PA announcement chime *Indicatiff Roissy*, composed by Bernard Paregiani, which was notably used to inform passengers at Charles de Gaulle airport, Paris, from 1971 to 2005, and which featured in the Roman Polanski film *Frantic*. This sensitised me to other PA systems that I was routinely encountering such as those in London (the Underground and rail network PAs) as well as the sounds in the Barcelona Metro. Incidentally, many public PA sounds from around the world are archived and available on the Web.

So, on reflection, the term 'technological soundscape' seems an apt way to capture how we, as a team, were thinking about the auditory qualities and characteristics of the Babble. At the same time, the notion of 'atmosphere' (I was only very vaguely aware of the German philosopher Peter Sloterdijk's use of the term and certainly not pursuing it) was being used to reflect on the emerging design of the Babble enclosure and glass chimney feature, and so we also started to consider the atmosphere of sounds the Babble might produce and how it might give rise to its own peculiar auditory environment or soundscape.

Back to the business of producing individual sounds or jingles. This work began with exploring and investigating existing jingles, from talk shows and PA systems, and using Logic Pro to transpose samples into MIDI to get a sense of the kinds of temporality, structure, and rhythms used. I also played around with taking longer

musical sections and extracting small fragments from them, which, in some cases, led to some 'lighthearted' staccato phrases played through software wind instruments, such as clarinets, which feature in the Babble. I also experimented with sampling or taking short phrases, and building up layered and concatenated repetitions through the sequencing interface, making longer sequences, and then cutting or extracting short jingle-length passages from these.

At the time, sequencing, layering, and phasing brought to mind the work of Steve Reich, which, admittedly, I've admired for some time. Not only was I reminded of Reich's pioneering usage of sampling (in the pieces *It's Gonna Rain* and *Different Trains* for instance), but also of the processual patternings of harmonies, which his music accomplishes through the intricate and disciplined interplay of instruments and musicians. In his paper 'Heterogeneities', John Law (2003) lets Reich's work, amongst that of other minimalist composers, speak to an actor-network theory sensibility, and in doing so considers 'social', material and political processes as ironically incomplete, unstable, and displacing as well as cumulative – music and the social are both seen as 'gradual' and literal processes (Reich 1974: 9). On a more practical note, my admiration for Reich's work found its way into the Babble sound design. Here, the changes between sections in *Music for 18 Musicians* that are cued by a metallaphone sequence (Reich and Hillier 2002: 90) have been directly transposed into the Babble as a jingle that segues into the announcement of SMS messages sent to the Babble as well as messages spoken into the Babble's microphone.

The soundscape of the Babble also, and very obliquely, alludes to Afrofuturist music, specifically a refrain transposed from Sun Ra's *Space Is the Place*. Placed alongside Sun Ra's engagement with racial politics – which his music clearly bears upon and is celebrated for

– in the context of the Babble this refrain can also speak to tropes of technological futures, space, and alternate histories.

Being part of a team designing an experimental device also took me to exploring other forms of experimental music-making devices. Incidentally, in his studio in 1969, Reich recounts abandoning his Phase Shifting Pulse Gate, somewhat early on in his career, citing the machine-like precision of the electronic prototype (1974: 25), which denied the micro-variations and their processual possibilities present in human-instrument configurations. In contrast, Raymond Scott's Manhattan Research project suggested that electronic devices have their own auditory and musical possibilities, and could also be playful, echoing Tati's incidental music. Scott's Manhattan electronic music also includes advertising jingles and film soundtracks.

The jingles that were designed for the Babble also feature in the system's built-in auditory user guide. When the volume knob is dialled counter-clockwise to the 'info' marker, below volume 0, the Babble describes what it does and how to use it, using the voices associated with the different kinds of internet source content it collates. This mode and each voice is cued by way of a jingle – much like the Indicatiff Roissy. In mediating as much of the interface as possible through sound, we also decided to make the Babble speak its volume setting as it is dialled in – switching to volume five, for example, is confirmed by 'volume five' being spoken by the system. Invariably, this meant doing various tests to ascertain the most appropriate volume levels as well as pauses so as to avoid overlapping read-outs when the knob is being turned through volume levels.

I appreciate that parts of my discussion above most likely come across as conceited and pretentious for what, after all, are simply pre-recorded jingles that are repeatedly triggered by particular commands.

Please bear with me, though, since the soundscape of the Babble, its physical form factor, and the spoken content are all 'playing' with techno-natures, broadly put: the idiom of energy-consuming appliances, the language of environmental politics as well as reports on environmental practices. In this light, the Babble is *facetious* in that it is explicitly inviting energy communities to re-engage with the issue of carbon reduction through nonsense, unfamiliarity, and alterity. In part, our gambit is that through alterity the issue of energy demand can be readdressed in different terms. It is, after all, a rather strange-looking and sounding device.

Lastly, and given the spirit of the ECDC project, it is worth mentioning the many proposals for the soundscaping of the Babble that remain on the cutting room floor. If we speak about the alternative energy-demand futures immanent to the Babble, then the Babble also has immanent design futures that didn't quite take hold. Bringing up possibilities passed or unmade is, perhaps, one way of materialising and concretising this immanence, so to speak, of design, if only through words. Taking inspiration from talk radio, we discussed and mocked up the inclusion of a gentle looping electronic 'night-time' motif that would continuously repeat during UK night-time hours, overlaid with sporadic content drawn from the day's content, calling to mind the much-admired Shipping Forecast on BBC Radio 4, and its evocation of distant and unsettled natures. A number of recorded environmental sounds were also tested out – a nod towards *musique concrète* – such as the sound of rain or thunder, but this sounded too representational, and too concrete. We also met a researcher from the Department of Music at Goldsmiths, and after much discussion and demonstration of a prototype system, he suggested adding microphones that could act as electromagnetic sensors, so that

the sound of the Babble would directly respond to its situated environs …

References

Latour, B., *We Have Never Been Modern* (Hemel Hempstead: Harvester Wheatsheaf, 1993).

Law J., 'Heterogeneities', <www.lancaster.ac.uk/fass/resources/sociology-online-papers/papers/law-heterogeneities.pdf > (Lancaster: Centre for Science Studies, Lancaster University, 2003). [accessed 7 July 2016]

Reich, S., *Writings about Music* (Halifax: The Press of the Nova Scotia College of Art and Design; New York: New York University Press, 1974).

Reich, S., and P. Hillier, *Writings on Music, 1965-2000* (New York and Oxford: Oxford University Press, 2002).

work progressed, it became apparent we were crafting a strange form of radio, one that could

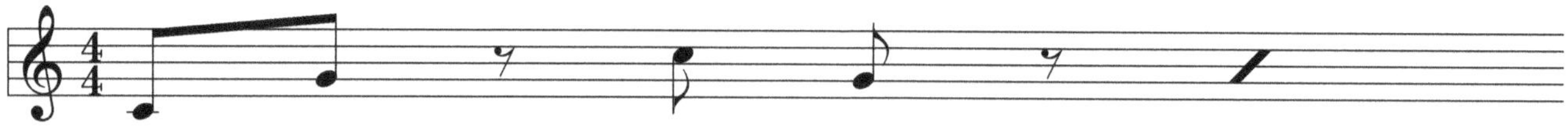

bAQ Clarinet

Babble Sound Stabs

Energy-Demand Chime

Hourly bSO Tick

Twitter bTE Chime

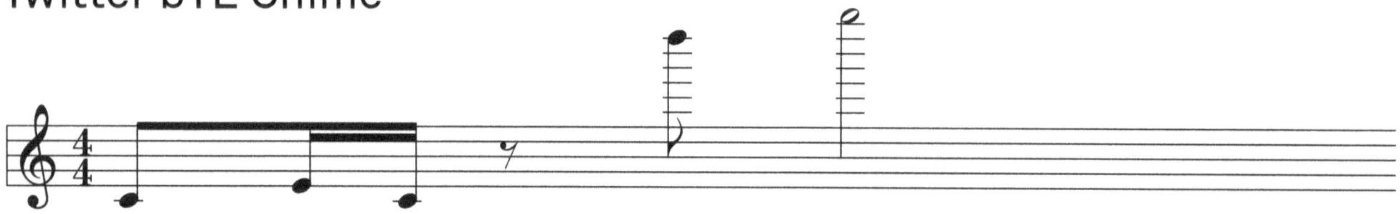

bLH Glock

**Different jingles were composed to introduce
each of the many voices of the Energy Babble**

be left playing continuously like an endless talk show. But where would we find all the

Software

Energy Babble software had to handle sending and receiving audio files between a server in the IRS and dozens of devices around the country. In addition, the system sent audio messages recorded on the Babbles to a Mechanical Turk speech-to-text service, and then merged the results with texts scraped from the Web for automatic speech synthesis. For the ECDC team, however, most of the buzz was around the 'bots' that trawled the Web – and particularly Twitter – to collect content for the Babble.

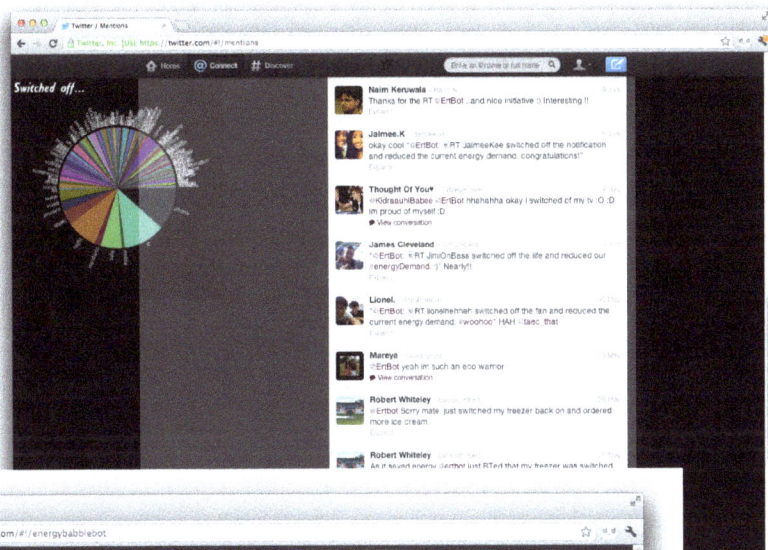

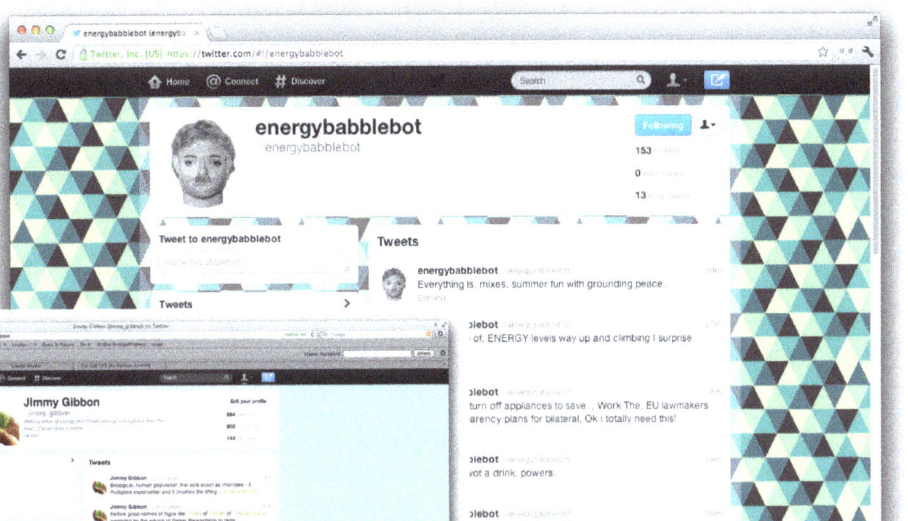

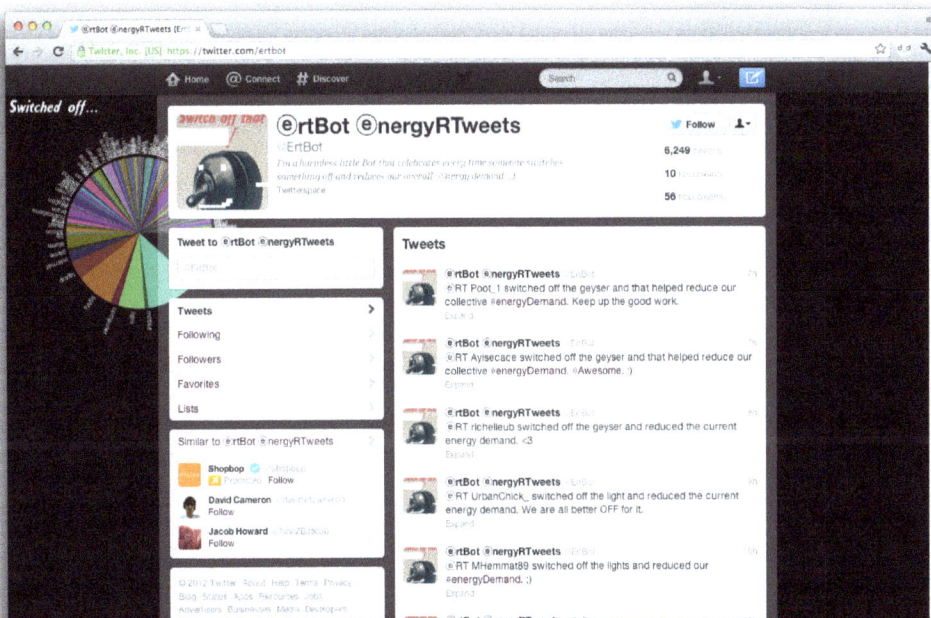

Early Twitter bots trialled a range of behaviours, retweeting eco advice (Eco Jo), celebrating objects being switched off (ErtBot), opining about energy matters (Energy Babble Bot) and commenting on current grid demand (UKGridBot).

Babblebot stories
Matthew Plummer-Fernandez

The Babble devices are brought to life by software which finds data and content online, and turns it into witty little stories about energy, which are read out in sound clips by machine voices. This software was generating material even before the Babble devices were made, as we wanted to engage an audience with the content that was being produced before we signed off the final system. To do this, we built a series of Twitter 'bots' that would send the automatically generated stories back into the world tagged with eco-sounding hashtags.

Bots are semi-autonomous internet robots, originally designed to carry out mundane processes of data gathering, but which became increasingly sociable and enchanting as social media channels enabled software developers to write applications that accessed the platform through the 'Application Programming Interface', or API. Twitter, in particular, became a testing site for creative developers to automate accounts, and try various techniques and approaches ranging from deliberately faking human profiles to more inventive uses such as spelling checkers that found other people's spelling mistakes and responded with the correction. Twitter bots are very economical and quick to make — you simply create a new Twitter account, register your application to retrieve the API keys, and use one of the many Twitter API software modules that allow developers to get their software to communicate directly to Twitter. Within a day, you can easily have a Twitter bot up and running.

When I joined the Interaction Research Studio I had only just started to discover bots. I got into using Twitter precisely to find bots and learn from the different strategies that bot creators were devising. I set up the blog botology.tumblr.com to document them all; sadly, the blog has lost its images due to a change in Twitter's URL format. Twitter has introduced many changes since then, even to its API, which has had the consequence of stopping many bots from running. The blog has become a time capsule of the Twitter bot Dawn Age; many of their actions are now prohibited, such as sending unsolicited responses to people's tweets. This is to curtail nuisance twitter automation such as targeted advertising. During the early era of Twitter bots, a popular strategy for creating original utterances was to use a Markov algorithm. This algorithm can be used to process a body of text and find probabilities of which word is likely to follow a given word. So, starting with the word 'if', you may obtain the next word as 'the', using the algorithm, which could be followed by 'weather' and so on, constructing a sentence such as 'If the weather improves I'll go outside'. The sentences generated are very much dependant on the text with which the bot has been trained. The @Fakespearean bot is typical of this type: it generates tweets from a Markov model trained on Shakespeare's complete works.

Markov algorithm-driven bots have become an increasingly unpopular bot tactic because of the nonsense they can create, but at the time the Studio and I were driven to explore this approach, and created our first bot @ecojo1. 'Eco Jo' was conjectured to be an environmentally conscious person sharing tips and retweeting tweets tagged with relevant keywords such as '#sustainability' and '#green'. Eco Jo's Markov-driven comments were trained on online guides that promoted sustainable living, resulting in tweets such as 'Make Your Own Products: Cleaners, Toothpaste, Shampoo I love making my own non-toxic cleaners. Think about it'. The bot managed to attract people with similar interests and this particular tweet got the response '@ecojo1 Now this gets my attention ESP the cleaner'. Through testing the bot, we could see that its success was very much dependant on creating a high number of tweets, as much of what it generated was garbage. Sometimes the web-sourcing would have accidently pulled in adverts, making Eco Jo say things like 'Amazon Price: $ 0. 99 5 .Wear Clothes More Than Once No, you don't have to wear your plaid shirt two days in a row'.

The Markov strategy made its way into the final Babble prototype as a sort of nonsensical character that could be heard coming from the device. The content was procedurally collected from articles about sustain-ability, as well as Twitter itself. The result was a very unpredictable nonsense generator which required a high degree of tolerance. The voice helped capture the tone of the conversations surrounding environmental issues, without having a particular message to convey. I personally see it as a form of abstract poetry, where meaning is purposefully erased to highlight the overall shape and colour of the dialogue. We can perceive it as 'green' without having to be told any particular green message.

Other Twitter bots exploited another strategy: mining numerical data and turning it into human readable phrases. @twrbrdg_itself, simply known as 'Tower Bridge', was one such bot, made by creative technologistTom Armitage to read out the activitiesof London's Tower Bridge. A typical tweet would say something like 'I am opening for the William B, which is passing upstream'. It did this

by sourcing the Bridge's schedule online, and tweeting its activity at the times when the opening and closing of the bridge was due to take place. Using this strategy, we created @ UKGridBot, which would scrape real-time information about the energy demand on the UK National Grid and convert
it into tweets such as '14:16 ▮ now: 46118 MW ▮ up by: 296MW ▮ GOING UP!!!'

The bot would later inform a more sophisticated version which would read out not only the total draw on the grid, but also its energy mix, consisting of energy drawn from nuclear, wind, coal, and other sources.

@ErtBot tested out another strategy for us, finding any mention on Twitter of things being switched off, and respond with a congratulatory tweet for reducing energy consumption. The bot would not be allowed to do this under Twitter's revised guidelines for automation,

but at the time was free to do so, and contacted many unsuspecting users. One particular example comes to mind from a user who was annoyed that someone had switched off the bathroom light whilst she was on the toilet. Ertbot's automated response only aggravated the Twitter user's fury, and she made it quite clear she was by no means interested in reducing national energy consumption during her ordeal. Ertbot's strategy made it into the final Babble prototype in a much less provocative capacity, by simply reading out once an hour all the things that Twitter users had tweeted that they had switched off. It would say things like 'three fans, two heaters, one iPhone ...', and so on.

Twitter bots have flourished in the last few years, and there is now a whole global community of 'bot makers' sharing strategies, themes, ideas, and each other's code. Some bots can amass tens of thousands of followers in just weeks. The Markov

strategy has declined, and is now considered a rather crude approach to creating interesting dialogue, and more controlled pre-worded templates and half-complete sentences are commonly used. Through the exercise of making the Babble, we uncovered that Twitter bots make excellent prototyping tools. Rather than waiting for the full completion of an interactive technology to reach its intended audience, the most fundamental engagement strategies can be tested on online audiences quickly and cheaply. The Twitter audience may not be anything like your target audience, and there are many other caveats to consider, such as the context and experience of Twitter compared with the context in which your research is due to operate. Nevertheless, nothing is lost in obtaining quick engagements with potentially hundreds of users to initially gauge the quality of your ideas for interaction.

(a13) algorithm model for a keyword-responsive, Learning 'voice' with listener's feedback and choice of keyword

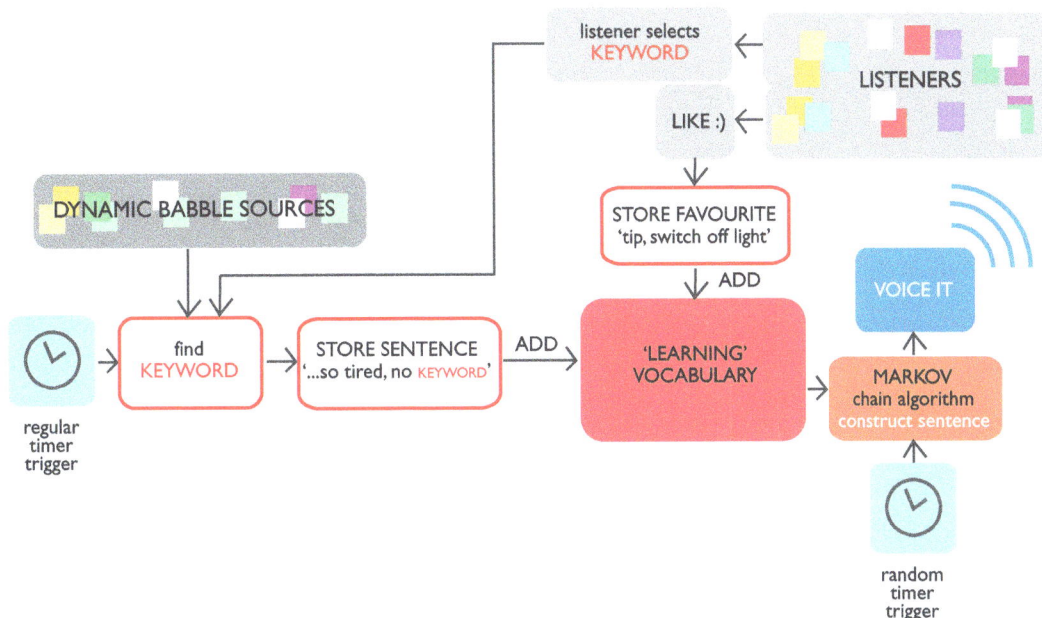

A series of algorithms and variables was developed to generate different interactions between users and audio content.

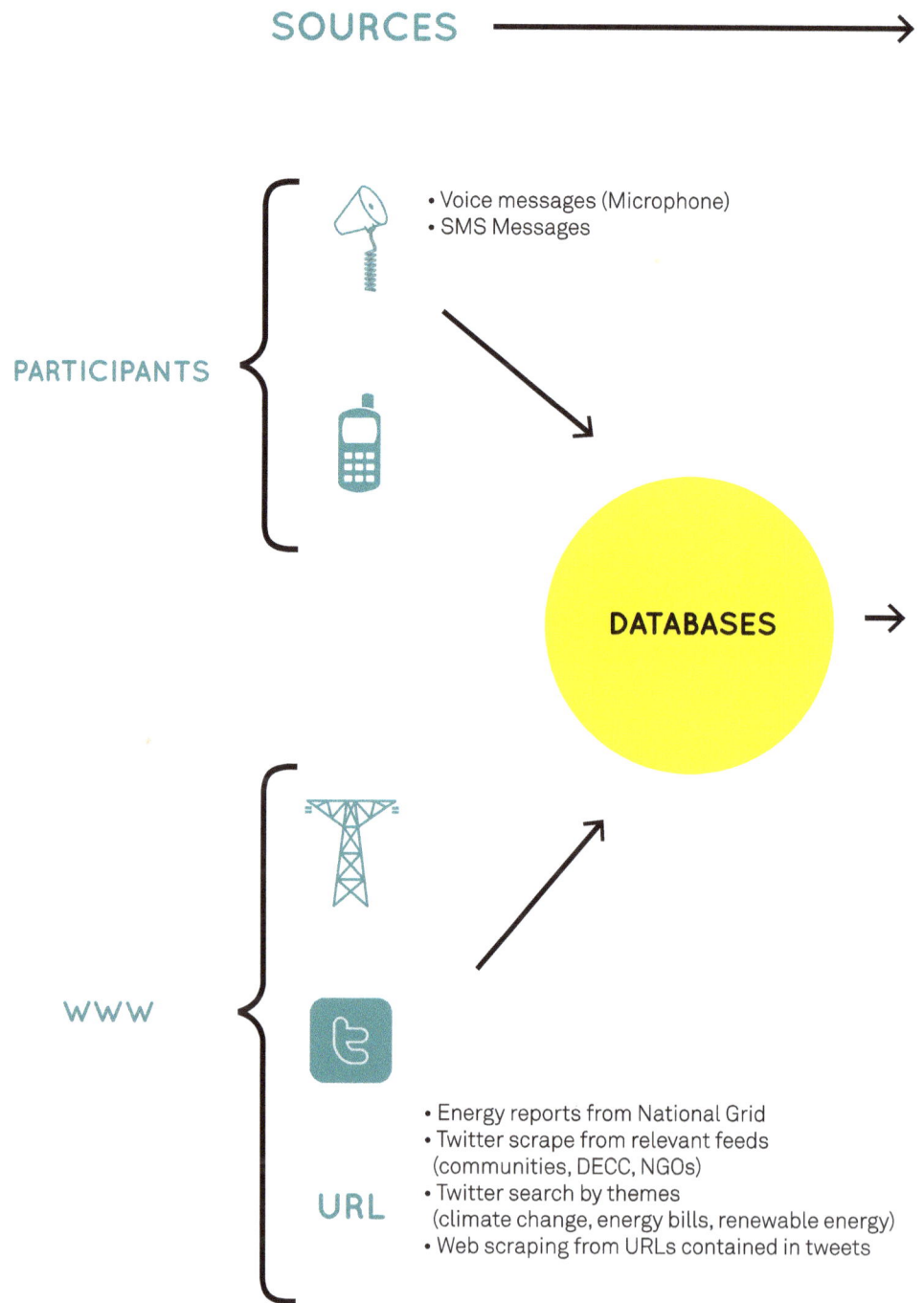

SOURCES

PARTICIPANTS

- Voice messages (Microphone)
- SMS Messages

DATABASES

WWW

URL

- Energy reports from National Grid
- Twitter scrape from relevant feeds
  (communities, DECC, NGOs)
- Twitter search by themes
  (climate change, energy bills, renewable energy)
- Web scraping from URLs contained in tweets

playful interventions, sometimes revisiting sites such as the National Grid for updates, and

ALGORITHMS  DEVICES

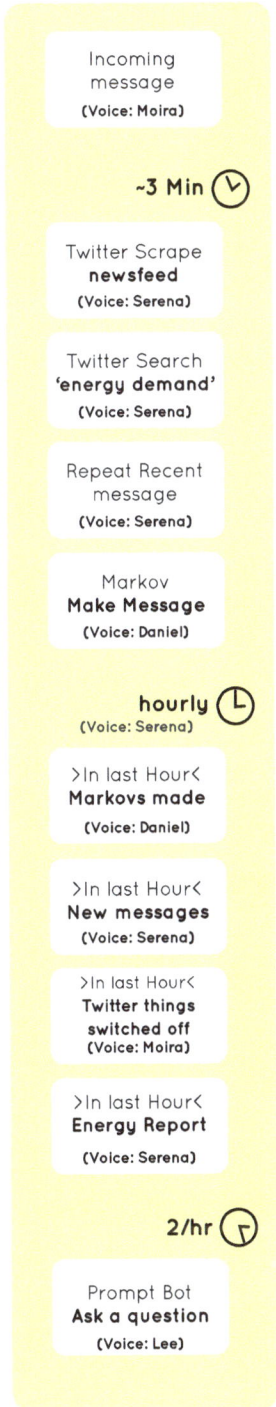

Incoming
message
(Voice: Moira)

~3 Min

Twitter Scrape
newsfeed
(Voice: Serena)

Twitter Search
'energy demand'
(Voice: Serena)

Repeat Recent
message
(Voice: Serena)

Markov
Make Message
(Voice: Daniel)

hourly
(Voice: Serena)

>In last Hour<
Markovs made
(Voice: Daniel)

>In last Hour<
New messages
(Voice: Serena)

>In last Hour<
**Twitter things
switched off**
(Voice: Moira)

>In last Hour<
Energy Report
(Voice: Serena)

2/hr

Prompt Bot
Ask a question
(Voice: Lee)

AUDIO STREAM
AGOLRITHMICALLY
CONSTRUCTED

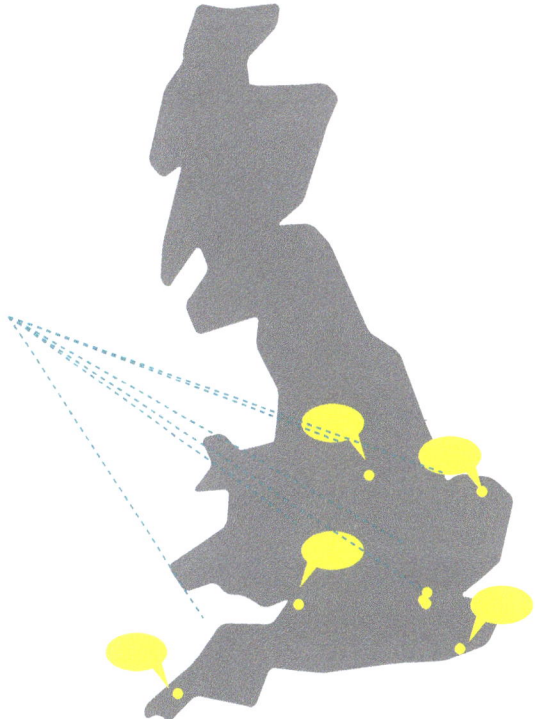

**Diagram of the algorithms and sources that
construct the Energy Babble audio stream.**

Form

The design of the enclosure for the Energy Babble was the result of a wide range of considerations and material exploration. System requirements, production, and functionality were entwined with aesthetic intent throughout the process of shaping the devices. To develop an aesthetic language, we looked into a variety of objects that would relate to the ambiguous nature of the audio system. From Braun radios, to bird nests and moiré patterns, these objects were used as resources to explore, sketch and prototype different possibilities for the material dimensions of the Energy Babble.

SK2 Radio by Braun

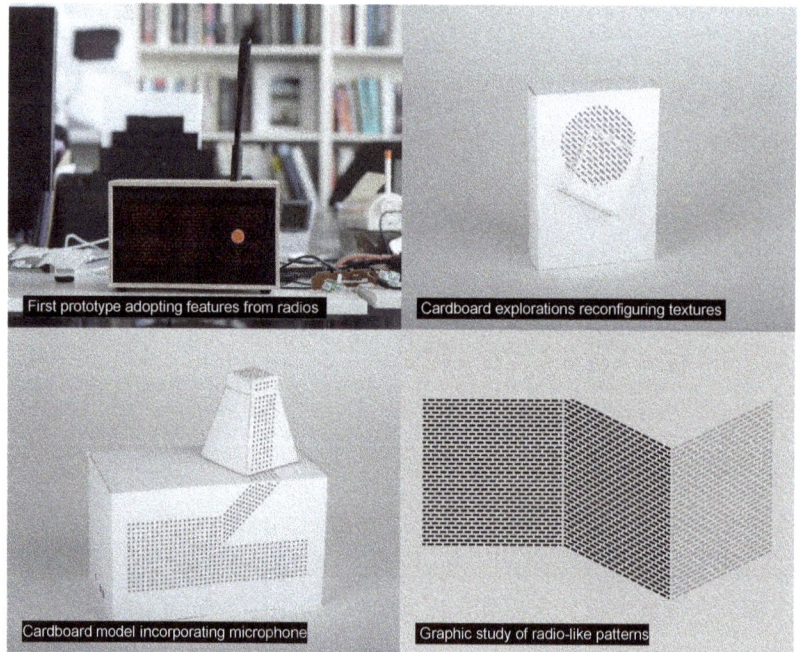

First prototype adopting features from radios

Cardboard explorations reconfiguring textures

Cardboard model incorporating microphone

Graphic study of radio-like patterns

Design exercises interpreting the SK2 Radio by Arthur Braun and Fritz Eichler designed in 1955. The diagram shows the first version of the device adopting radio features and further reconfigurations of speaker patterns.

Meanwhile, we started exploring the form design for the devices we would produce. The

Burrow

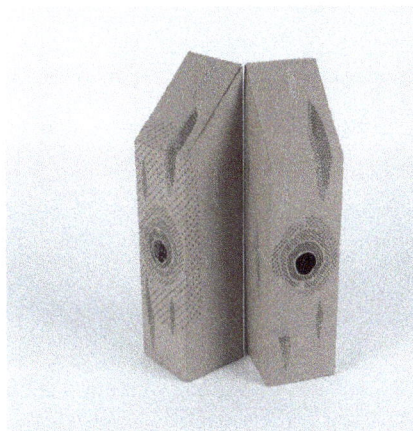

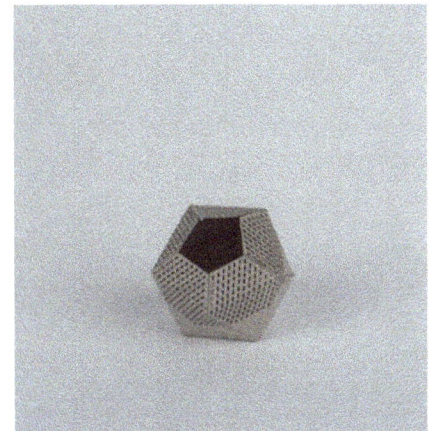

Laser-cut cardboard models of enclosures based on burrows and nests

By referring to objects that are not directly related to audio devices, such as burrows and nests, we opened a different area of exploration. The laser-cut models show interpretations of irregular textures applied as speaker grills.

Cube with ambiguous space by Jesus Rafael Soto

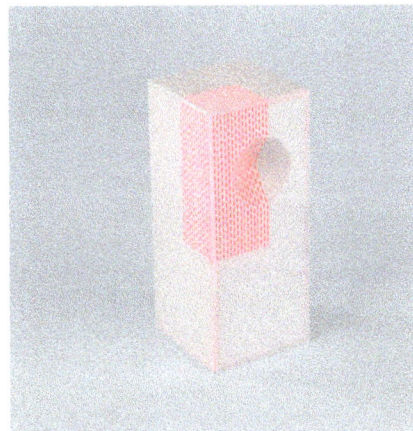

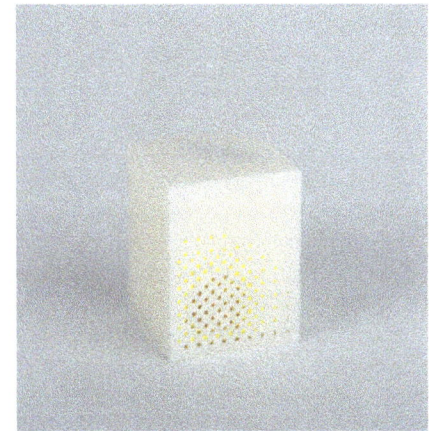

3d print models exploring optical effects by applying moiré patterns

To understand more about moiré patterns we referred to different examples of Op Art such as 'Cube with Ambiguous Space' by Jesús Rafael Soto, shown on the left. These observations were developed into 3D-printed models that played with the optical effect.

Shaping the Energy Babble

Liliana Ovalle

The Interaction Research Studio designs, builds, and deploys devices into the lives of people who volunteer to participate in our research. We do this both to demonstrate new technical possibilities and as a way of investigating how people engage with technologies and topics in their everyday lives. Because many of our devices are noninstrumental and open-ended in the ways in which they draw attention to issues, we believe that letting people live with them as they do with 'real' commercial products is the only way to find out what they mean.

This requires designing for a variety of social settings, which have included domestic spaces, elderly care homes, a remote island, and a convent. A key question the Studio encounters when developing a system is what kind of 'thing' is the artefact going to be? In other words, what are the aesthetic and material qualities of the device, and how will they influence the way it settles in to the environment for which it was designed?

When a novel artefact enters the everyday life of a volunteer, it takes part in a process of domestication (Silverstone and Haddon 1996). Whether the form is designed to have an expressive character or to be a simple black box, the elements that shape the object's appearance, such as proportions, materials, and colours, will prompt interpretations that may influence the situations in which it will take part. In our practice, we carefully consider the material embodiment of the interactive systems we design as their aesthetic attributes can contribute to helping people understand what they are and how they might be approached. This article will describe some of the material practices, insights, and decisions encountered throughout the design process of the Energy Babble as a material thing, with the aim of unfolding the design complexity encountered while developing an aesthetic approach to a novel object.

As an audio-based research device, the Energy Babble is designed to engage its users – members of local communities – in a playful and ambiguous way to reflect on issues relating to energy in order to reframe the problem of energy-demand reduction. This ludic design approach provided an important framework for aesthetic explorations in that the physical presence of the object would contribute to setting the tone of the engagement. If we think of Madeleine Akrich's notion of script (Akrich 1992) as the set visions and scenarios that designers 'inscribe' into a new object, the script that would define the Energy Babble's body would be one to prompt curiosity and openness rather than prescribe explicit forms of engagement.

With this agenda in mind, we started to work on the design of the Babble as a thing. By this point, the experiments with software and hardware were already in progress and the experience prototype (see section on 'Hardware' within 'Designing') had given us an idea of the nature of the algorithmic voices and their ambiguous content. In parallel, the sound design was evolving, introducing playful jingles to the soundscape. As the peculiar broadcast of the Babble was coming together, the design of the enclosure began to cover ground in different directions. In what follows, this article recounts specific steps of this process, highlighting the key moments that influenced the development of the aesthetic and physical attributes of the device.

Setting the ground: exploring design spaces

Early engagements with the project's participants using Cultural Probes, design workbooks (see 'Cultural probes' within 'Framing'), and field visits produced insights that helped shape a framework of exercises for the Energy Babble. These materials not only informed the main concept for the device, but they also allowed us to draw references to existing objects that could help people relate to the Energy Babble as a 'thing', including energy monitors, confessionals, radios, and megaphones. This diverse selection of objects provided a space to start the aesthetic explorations, allowing new design directions for the form to emerge.

Each of the design directions described below was driven by a particular intention, and an important part of the design process consisted in expanding the design space by introducing diverse objects and cultural references that would help us to draw the material features of particular shapes, textures, and proportions that were interpreted, prototyped, and reflected upon.

Incorporating such diverse materials into the process allowed us to construct an idiosyncratic space to design the aesthetic of the Energy Babble. For the purposes of this article, the directions are classed in five categories; but it is important to note that in practice these areas of

the concepts behind it and to help shape its identity. Our first studies sprang from

exploration evolved organically and in a non-linear fashion. Each iteration was informed by the prior experiments, and the results would add to the chain of considerations and decision-making that led to the material language of the final design.

1. Design space: the non-aesthetic body

As a preliminary exercise in the design of the Energy Babble enclosure, we produced a small batch of experience prototypes of the system (see 'Form' within 'Designing') that were deployed amongst members of the team. The aim of this trial was to test domestic installation of the system and to have a closer experience of broadcast from the Babble by bringing it to our own domestic environments. The prototype included the essential functional aspects of the system, consisting of a Raspberry Pi housed in a laser-cut case, a USB speaker, a USB microphone, and an ethernet cable for connectivity. The case, made of Perspex – a material commonly used amongst laser-cut products – was adapted from an open-source design to include a button that allowed the microphone to be operated.

The enclosure of this prototype did not intend to explore any particular aesthetic direction, but simply to house the minimum elements required for this version of the system to work. We could think of the resulting object as a non-aesthetic body, it presented itself as a hybrid assemblage of electronic components. While this version of the Babble's body does not address a ludic approach, it allowed us to understand the behaviour of the system, its polyphonic and opaque character as well as the flow of the broadcasts. As the technological platform evolved to its final configuration, the number of custom-made and off-the-shelf components grew in number and complexity to fulfil the system and connectivity requirements. The aesthetic explorations that followed had to adapt to and embrace the new specifications.

2. Design space: thinking of radios

Since we were designing a broadcasting device, a natural starting point for shaping the Energy Babble was to look at the material language of radio receivers. We collected images of radios from different eras, from Philco radios from the late 1920s, to deconstructed radios by Daniel Weill in the 80s. Most influential in these explorations was the work by designers Dieter Rams and Arthur Braun, particularly the ss SK2 and the RT20 radios developed and produced in the 60s. These iconic machines responded to Rams' ethos of 'good design' which still prevails in the design of electronic appliances, in particular in current Apple products.

We developed different iterations based on these radios, in which the Babble adopted some of their key features, such as perforated surfaces for speaker grills, controls, and proportions. However, the result of this exercise presented a strong resemblance to existing radios, providing too specific a frame for the experience of the new system. To distance the object from direct radio references we decided to do further explorations by disrupting some of the elements.

One relevant exercise consisted in treating perforated speaker grills as a texture that could be reconfigured and would still echo the relationship to audio appliances (see diagram on page 67).

3. Design space: opacity of broadcast sources

Other concepts for the Babble's aesthetics emerged by reflecting on the nature of the flow and transparency of the content it broadcasts. The Energy Babble gathers content from different sources including Twitter and direct contributions from users; however, the origin of this content is not always transparent. While some content is processed to remain anonymous, some is mixed and rephrased by a series of bots implemented in the system, resulting in a stream of ambivalent and sometimes nonsensical data.

With this is mind, we began exploring the concept of opacity. Initially, we took references from burrows and nests, thinking about how such concealed spaces can trigger curiosity, and we began applying concepts of hidden spaces to maquettes of possible enclosures. This exercise manifested in multiple laser-cut cardboard models that reinterpreted the organic housings into geometrical shapes with perforated textures. These explorations, along with the radio exercises described above, allowed us to discover the potential use of moiré patterns in the enclosure. In both cases, overlapping laser-cut patterns created undefined visual spaces. To understand more about this effect, we referred to different examples of Op Art, such as the work of Carlos Cruz-Diez and Jesús Rafael Soto, includingg 'Cube with Ambiguous Space'. These observations were developed into 3D-printed models that played with the optical effect (see diagrams on page 67).

4. Design space: a vessel for a soundscape

As the proposals were evolving, we encountered technical difficulties in managing the quality of the audio. Most of our designs relied on the idea of enclosing a ready-made portable USB speaker, and transmitting the sound through a perforated surface. But the prototypes demonstrated that this format decreased the sound quality. To expand our options, we looked into other types of audio artefacts, musical instruments, and sound installations. At this stage, the reference of the megaphone re-emerged from previous workbooks, which related the device to forms of activism. The megaphone became a crucial element for the final design both in shape and functionality. This switch of direction also clarified the concept of the device as a container of a soundscape. In this configuration, the megaphone would be the extension to magnify and deliver the audio content to users. To apply this principle, we sketched and prototyped different enclosures with single- and multiple-audio outputs (see diagram on top page 71).

5. Design space: capturing and delivering voices

After deciding that the system would use a megaphone to deliver the soundscape, we explored different ways to

classic radio designs, but we quickly moved towards expressions of enclosure and release,

resolve the object aesthetically and functionally. The conical shape of the megaphone not only delivered the sound of the Babble's voices clearly and legibly, but it also provided a strong visual indication of a vocal sound output. Based on this, we defined the shape of the microphone as an inverted cone or funnel as a way of indicating content input. While researching materials for the megaphone, we discovered that glass has excellent qualities for amplifying sound. To incorporate this material into the enclosure of the Babble we looked into laboratory glassware. Objects such as flasks, beakers, and funnels were used as a reference to further develop the glass features of the device (see diagram on bottom page 71).

Towards a final design

These design exercises gave us an understanding of the material language that would define the Energy Babble as an object. The product development that followed this stage was continuously informed by the insights and discoveries described in each direction explored. The sketches, models, and tests allowed us to develop the specifications, affordances, and further forms that shaped the object as a whole (see pages 88–89). The final design consists of a combination of two custom-blown glass components set on top of an injection-moulded base. The central glass piece works as a megaphone that amplifies the sound output of the device. This piece is screen printed with a pattern that alludes to the speaker textures explored in the radio exercises and moiré effects. The second glass piece

enfolds the base, the megaphone, and a plastic microphone together, while securing the elements in place. The electronic components and speaker are enclosed in the plastic base, which also provides a visual frame for the composition of the different elements.

By describing the different design directions, we experimented with during the embodiment of the Energy Babble, I highlight use of cultural references and existing typologies as a resource to ground the aesthetic explorations when designing novel artefacts. These references can provide a range of directions for materialising a system; however, they are not to be read as fixed meanings or explicit metaphors, but as suggestive media that will give a sense of character to an unfamiliar object.

The iterative process of materialising the device under different intentions allowed us to create a unique design space that led to the idiosyncratic aesthetic language of the research device. The value of these aesthetic properties in the context of a ludic engagement can be best understood by way of users' interpretations that emerged during the deployment of the device. Many participants remarked favourably on the physical form of the Energy Babble, referring to it as a 'nice, funky-looking thing', or admiring it as a 'lovely visual element'. Others were intrigued by it, describing it as a 'retro electric-appliance'. While it evokes a certain familiarity, the Energy Babble can be seen as an object that resists definition, inviting the user to form their own views around it.

References
Akrich, M., 'The De-scription of Technical Objects', in W. E. Bijker, and J. Law, Shaping Technology/Building Society: Studies in Sociotechnical Change (Cambridge, MA: MIT Press, 1992), pp. 205–24.

Silverstone, R., and L. Haddon, 'Design and the Domestication of ICTs: Technical Change and Everyday Life', in R. Mansell, and R. Silverstone, eds, Communication by Design. The Politics of Information and Communication Technologies (Oxford: Oxford University Press, 1996) pp. 44–74.

Image Sources
Intonarumori: Luigi Russolo, Instruments built by Russolo, The Art of Noise, 1913. (Retrieved October 4, 2015) https://commons.wikimedia.org/wiki/File:Russolointonorumori.jpg

Bag pipe: Ellsworth D. Foster ed. Bag pipe. The American Educator (vol. 1) Chicago, IL: Ralph Durham Company, 1921. (Retrieved November 23, 2014) http://etc.usf.edu/clipart/50200/50251/50251_bagpipe.htm

Organ Pipe Types: William Henry Stone, 1879. Elementary Lessons on Sound, MacMillan & Co., London, p.165, fig.56 (Retrieved November 23, 2014) https://commons.wikimedia.org/wiki/File:Organ_pipe_types.png#filelinks

Henley Megaphone (Retrieved March 30, 2015) http://www.nauticaliatrade-sales.com/henley-megaphonebrass-30cm-2141?filter_name=2141

Laboratory Cup and Jar. Elroy M. Avery. School Physics. New York: Sheldon and Company, 1895 (21). (Retrieved November 23, 2014) http://etc.usf.edu/clipart/20000/20029/cupandjar_20029.htm#

**Sketches exploring single and multiple
audio outputs resulted from the
interpretation of sound systems such
as Luigi Russolo's 'Intonarumori',
a bagpipe, organ, pipes and a megaphone.**

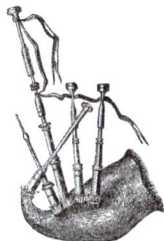

References of sound systems and instruments: Luigi Russolo's 'Intonarumori', a bagpipe,
a megaphone, and organ pipes

Sketches exploring single and multiple audio outputs

**Glass laboratory equipment such as
funnels, flasks, and jars provided
a reference to develop the glass
elements of the Energy Babble.**

Laboratory cup and jar

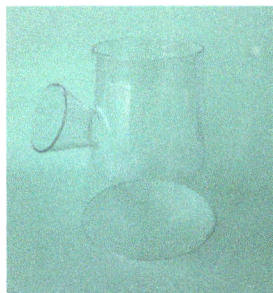

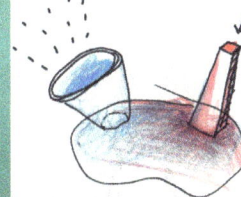

Models and sketches of glass shapes

Eventually we moved towards traditional loudspeaker cones and trumpets, and labware in

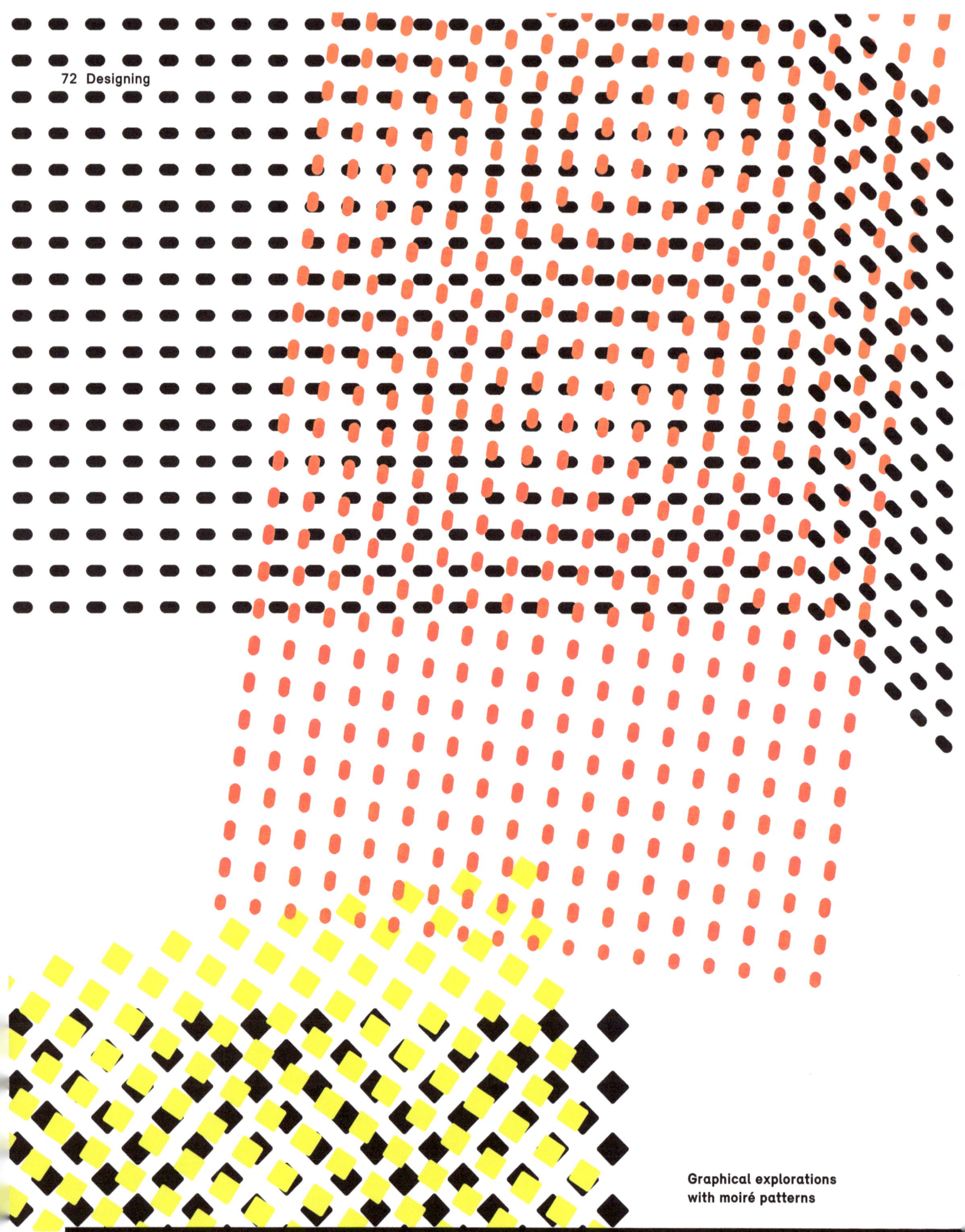

**Graphical explorations
with moiré patterns**

particular, because glass is an excellent material for speakers. As the design came together,

we started turning our efforts towards developing the Babbles for batch production, so that

Production

The design of the Babble exploits a diverse range of manufacturing techniques from the handcrafted to the latest 3D digital processes. Each part was fabricated to a high standard as appropriate for the batch production of thirty identical devices expected to run fault-free during a long-term field trial. The glass components of the enclosure are individually blown by makers who specialise in producing laboratory glassware. The plastic base and microphone were injection moulded by a company with expertise in low-volume production of plastic parts using an automated factory. The coiled microphone cable was formed by a cable specialist who wound the cord to the particular dimensions of our design. The Button and dial were 3D printed, as was an internal structure to hold the electronic hardware, which was a mix of off-the-shelf components and custom-made printed circuit boards. Once all the parts were produced, all thirty devices were carefully assembled at the Studio and prepared for deployment. Custom packaging and manuals were designed and produced to take the devices to their multiple destinations across the UK.

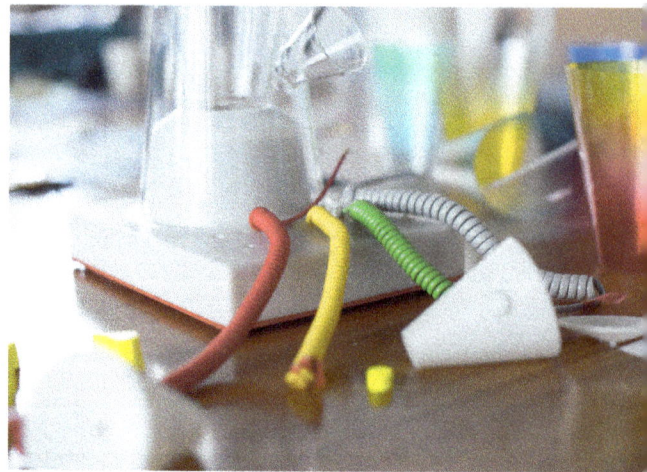

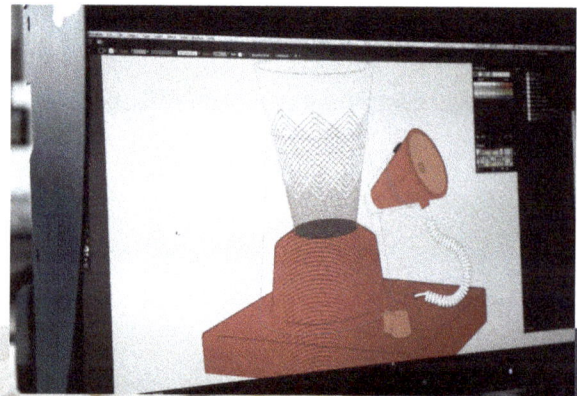

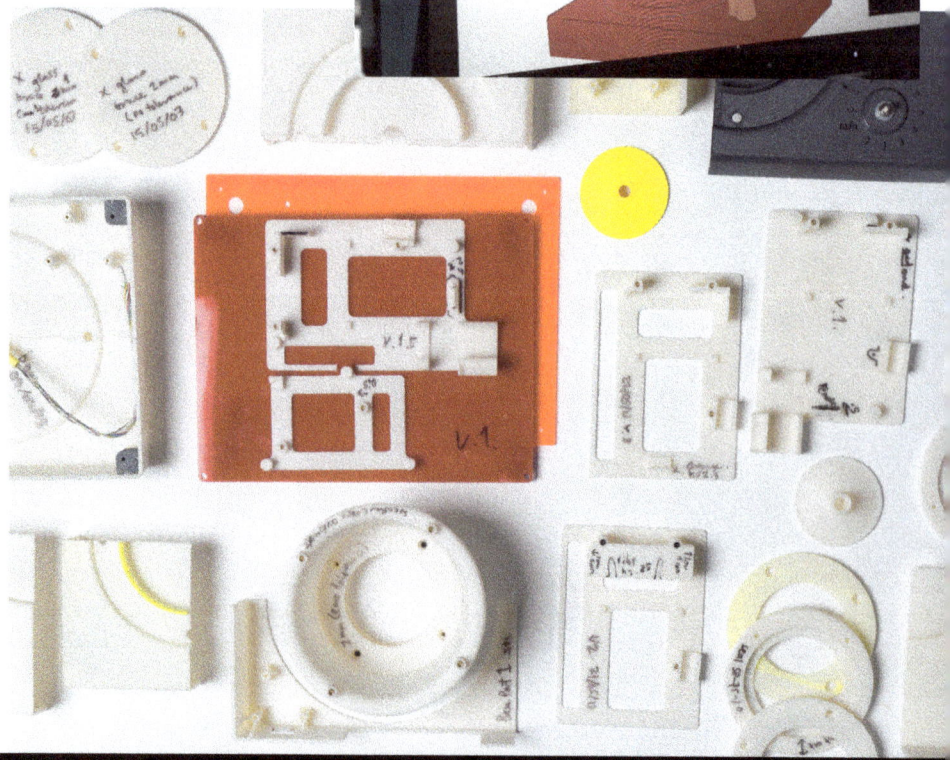

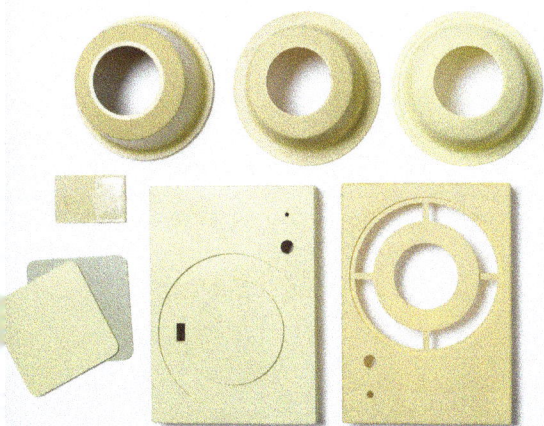

Multiple iterations of each component were developed: from coiled cables, and glass treatments to colour tests.

component of the devices, before assembling them in our studio. Since the rapid prototyping

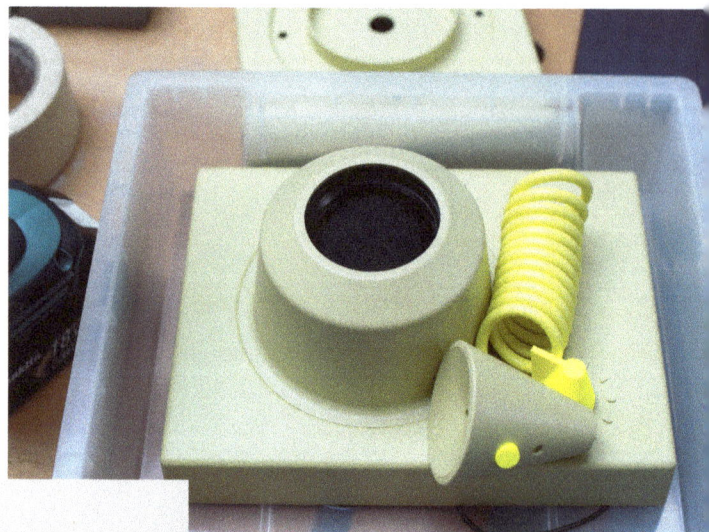

**Batch production and
assembly of the devices**

approach we had used in previous projects would be impractical, we decided to have the

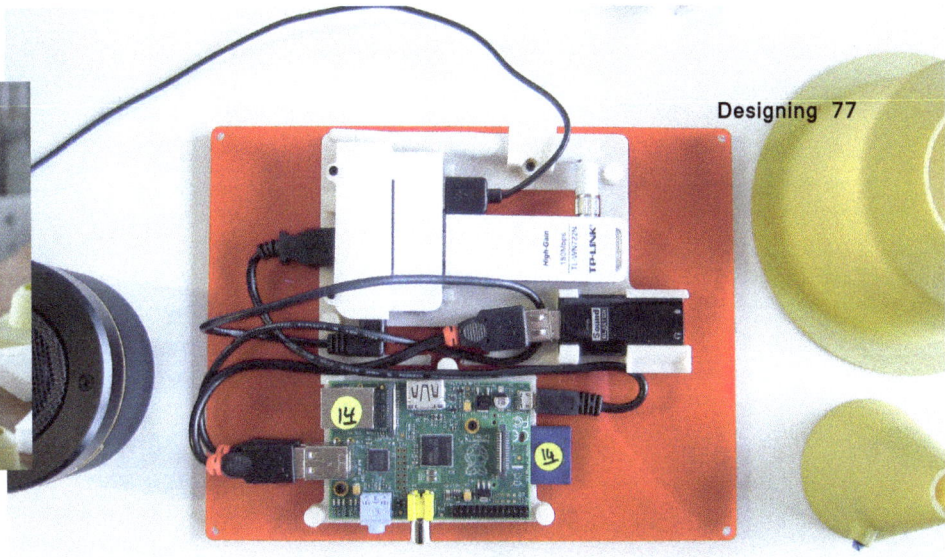

cases injection moulded, and to have the glass elements made by a South London glass-blowing

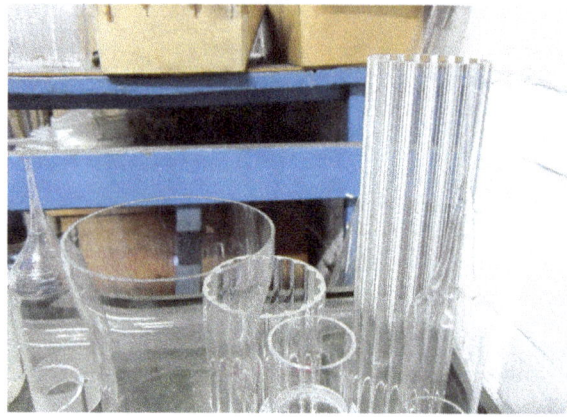

The glass megaphone, designed
to amplify the sound of the Energy
Babble, was produced at a local
manufacturer of laboratory glassware
using traditional techniques

factory (whose workers were overheard expressing sceptical incomprehension about what w

The microphone and the algorithm
Andy Boucher

I started working on the Babble prototype at the stage when it was being developed for batch production. At this point, the team had decided to have the fabrication of the electronic housing outsourced to a company that specialises in low-volume plastic injection moulding. This process is normally used in mass manufacturing to produce thousands or even millions of units. However, by automating workflows and introducing techniques that produce cheaper, shorter life tooling (the mould), the industry has made this process accessible to makers of low-volume artefacts. The advantage of plastic injection-moulded over handmade or 3D printed housing is that the components will be very strong and will require no post-processing; in addition, the plastic can be formed in any colour. Finishing and painting parts is labour-intensive in most other forms of manufacturing, and we were keen to avoid this step with the thirty-five Babbles that were planned for production.

The team looked at several companies offering low-volume plastic moulding services and decided to use a manufacturer called Protomould after being impressed on a factory visit at the quality of their finished parts. Like others in the industry, Protomould uses computerised machining as part of a package of methods that streamlines the process of producing tooling; but what made this company stand out was its automated quotation facility. For the customer, this is first encountered through their web-based service that uses software to analyse uploaded 3D drawings. Within minutes of submitting

a part, the system generates a full report of advice highlighting potential problems and improvements that could be made, as well as a complete breakdown of costs for tooling and unit prices for various scales of production. The speed and sophistication of this system meant that injection-moulding novices like us, could practise designing and engineering our 3D parts without wasting (or spending) time on expert human advice.

The reports that were emailed back consisted of three sections. First would be a price for the basic tooling cost (which would be the main fee) and the unit charge of production parts (which would be tiny). Second would be an illustrated section on the *required* changes, which advised on whether the submitted 3D form was actually mouldable. A viable drawing

would have no recommendations in this section, but it was surprising how often parts would fail this test. There are simple rules for moulding regarding draft angles and wall thickness, but ensuring that 3D forms drawn in virtual environments follow these rules can be tricky, and minor oversights are common. Third, the software provided advice on areas that might be problematic even if the part were inherently mouldable. These potential issues were presented as advisories, but could include problems such as excessive shrinkage where the wall was too thick, or areas that might collect drag marks when the part was ejected from the mould. There was, in fact, a whole host of potential issues that might lead to the finished part not being up to the expected quality. Many of these were predictable, but some were idiosyncratic to the Protomould system, which could only really be learnt by practice.

I took on the task of using this system to prepare the microphone case for injection moulding. The final form had already been designed, so my job was to work internally on creating space and fixing points for all the electronics, ribs to give strength and rigidity, open apertures for a button and cable, and holes in the mouthpiece

so sound could pass through to the microphone hardware. As always with moulding, the goal was to do this in as few parts as possible to save on the cost of tooling for each component. In simple terms, the microphone case was shaped like a stubby cone, with the voice aperture at the larger end shaped like a dish, curving into the main body. This detail, along with the need to mount a circuit board via a mechanical fixing inside the body, meant that the case ideally needed to be formed of three parts: two shell-like halves to create the cone shape and a shallow dish to form the microphone mouthpiece. It was possible to form the case from two parts, but this would have complicated the mounting of the button, and compromised the reliability of the device by providing inadequate means to anchor the cable, which we imagined would be under heavy strain during use.

I approached my first upload to the quotation system unaware of how sophisticated the advice would be, assuming that it would be biased only towards the economics of producing the parts. The first 3D drawing I submitted was simple: I split the microphone in two, and hollowed the solid forms into two shells; there were undercuts around the mouthpiece so I knew it was unmouldable, but I wanted to see how the system would respond. As expected, the obvious issue was highlighted, but the attention to detail was surprising: the system had scrutinised the part so thoroughly that every possible issue was exposed. A human would have thrown my careless drawing back at me, but the algorithm dutifully checked and highlighted every fault with a straight face. I have a reasonable amount of experience of creating plaster moulds for slip-cast ceramics, but there are idiosyncrasies to plastic (and in particular this system) that were unfamiliar, and so I had a steep learning curve over the next few uploads of more serious attempts at drawing the parts.

It took me a few attempts to pass all the tests of my algorithmic tutor,

but I began to learn an enormous amount about plastic moulding techniques. For instance, we had decided that the parts should be lightly textured, but this required a five-degree draft angle on any plane perpendicular to the tool's parting line. This was a surprisingly large amount, and the effect can be seen at the smaller end of the microphone case, which is far more pointed than Liliana Ovalle's original flat design. Second, wall thickness needs to be very carefully controlled. Areas that are too thick can cause deformations in the final part such as sink marks, shrinking and warping. However, areas that are too thin cause similar problems. Ensuring that the microphone case had a consistent three-millimetre-thick wall involved a large amount of detailed modification to internal corners and edges that would never be seen.

Once I had jumped these hurdles, and was getting clean reports, I could begin to focus on the overall cost. I began to see how adding a bit here, removing a bit there subtly affected the tooling price, which was largely dictated by the volume of material that would need to be machined out of an aluminium block. The shallow tool for the mouthpiece was a few hundred pounds, for example; but the deeper tools for each half of the cone were six times as much. As I examined the individual reports for each half of the cone, it occurred to me that they were so similar (let's call them 'part A' and 'part B') that a substantial saving could be made if the cone could instead be formed from two identical components.

There were two issues to overcome: first, the edge that forms the seam between the two parts had a groove on part A and a lip on part B, which is the standard way to lock together the edges of plastic shells. Second, part B had an additional hole for a single machine screw to mechanically lock the parts together. After some thought, I redesigned part A to have a lip around half of its edge

and a groove around the other half. These were oriented around a central axis so that when two part As were brought together, the edges would lock to form the cone. With a few additional alterations, I could eliminate the need to make a tool for part B. The apertures for the speaking button and cable were almost symmetrically aligned in each half of the cone, so with some minor tweaks these could both be formed by the same hole in the mirrored part. This only left the issue of the hole for the fixing screw. The most straightforward solution was to delete the hole from the tooling and instead drill it out of the bottom halves of the cones as a post-production step. Although we were trying to avoid any additional finishing, a drilling jig allowed us to complete this process quickly and accurately.

I would never have considered experimenting with ideas for saving on the tooling costs if it were not for the automated quoting system. The double-sided solution for the cone is probably not unique, and is of course only viable due to the symmetry of the form. However, the ability to rely on such tireless algorithmic advice to keep testing new concepts encourages such investigation. The finished microphone on the Babble looks pretty much like Liliana's original outline (except for a slightly more pointed narrow end) and is a small part of the overall device. It belies the effort required to batch-produce thirty-five of them and doesn't reveal the internal engineering that makes it work as a completely unique and robust prototype microphone used by so many participants. I have often described the Babble to others as a kind of automated talk radio station because of the complete dependence on automation and algorithms of its content — as well as its physical presence, if we consider the infrastructure behind its production.

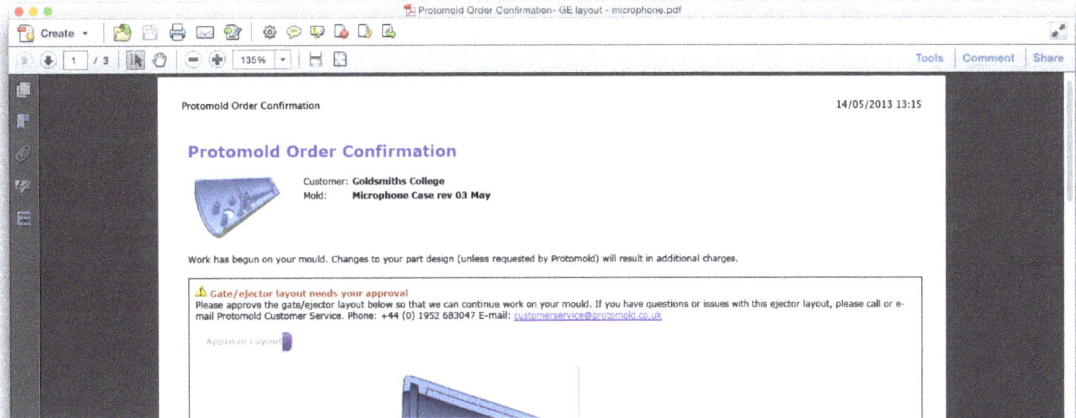

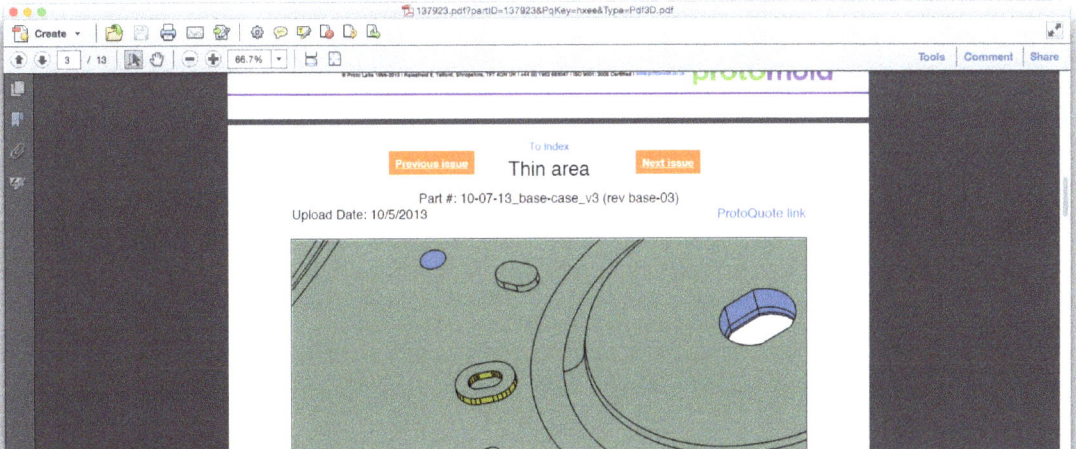

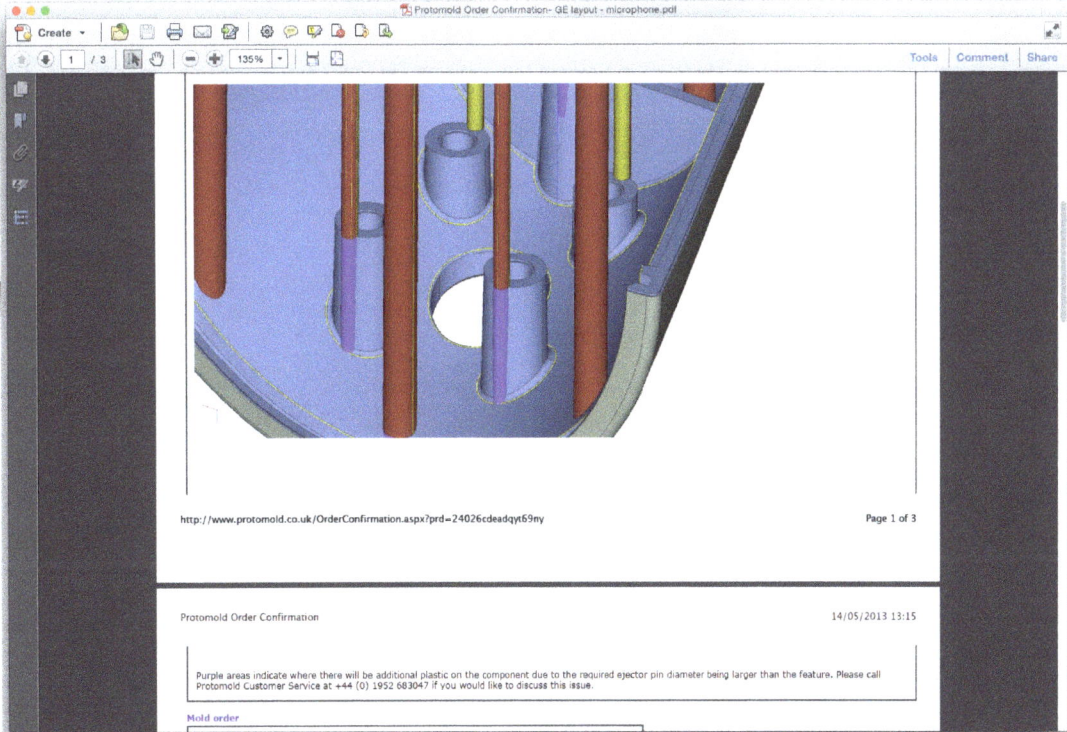

Recommendations and suggestions provided by the part analysis software of the plastic injection moulding manufacturer

with each of the elements of the devices subject to consideration, discussion, scrutiny, and

Exploded view showing all the components of the Energy Babble. The device contains off-the-shelf components, notably fixings and some of the electronics, and parts that we designed and built.

dispute. Parts started to arrive from the manufacturers, electronics warehouses, and our ow object printers, and we became a small-scale assembly house until we had finally constructed

A fully assembled Energy Babble device

35 fully functioning, completed Energy Babbles. We were very pleased with the results. But

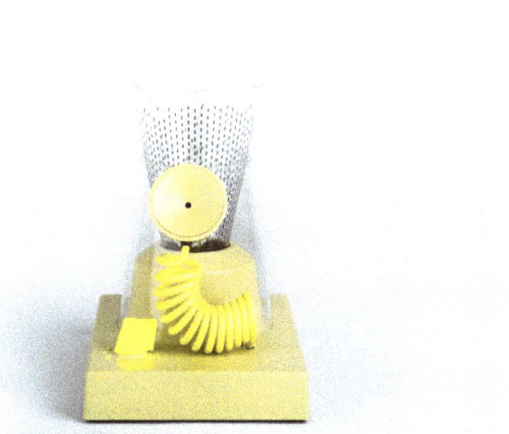

our work was not over: we still had to develop packaging, a user manual, and 'ethics forms'

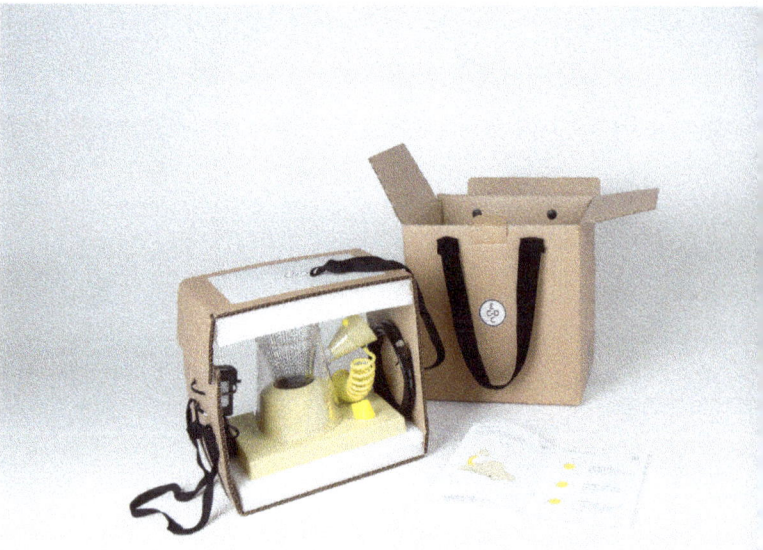

**Packaging was designed to
transport the Energy Babble to
different locations across the UK**

asking participants to consent to our collecting data – videos, photographs, quotes – abou

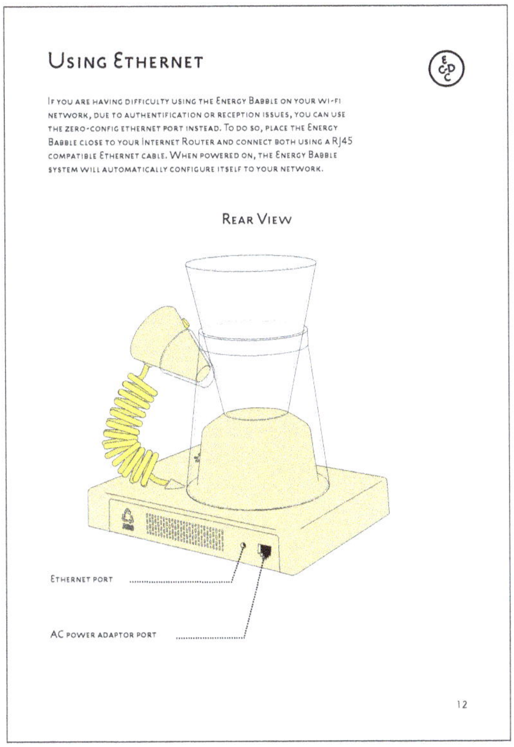

USING ETHERNET

If you are having difficulty using the Energy Babble on your Wi-Fi network, due to authentification or reception issues, you can use the zero-config Ethernet port instead. To do so, place the Energy Babble close to your Internet Router and connect both using a RJ45 compatible Ethernet cable. When powered on, the Energy Babble system will automatically configure itself to your network.

REAR VIEW

Ethernet port ....

AC power adaptor port ....

12

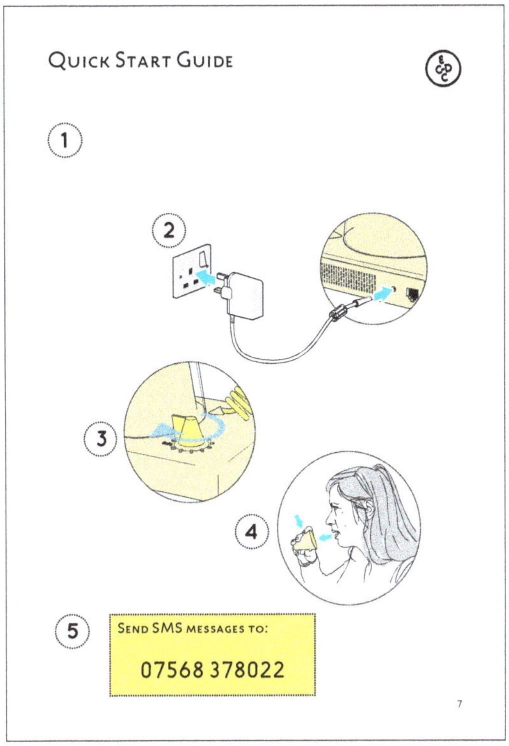

QUICK START GUIDE

1

2

3

4

5   **SEND SMS MESSAGES TO:**

07568 378022

7

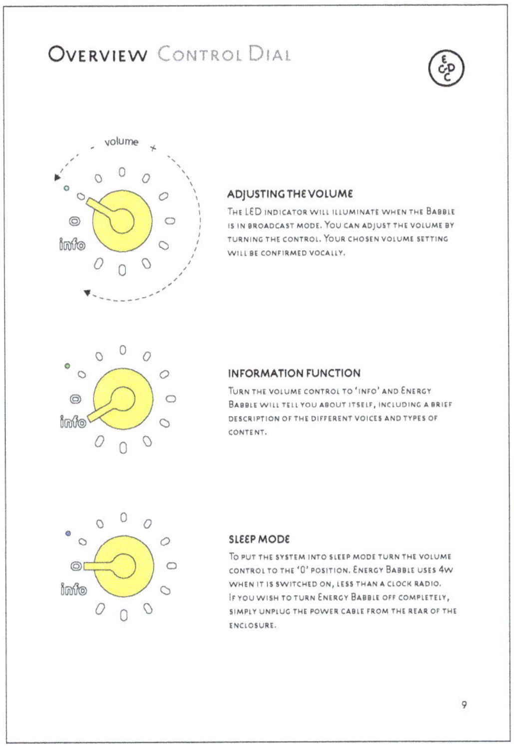

OVERVIEW CONTROL DIAL

volume

info

ADJUSTING THE VOLUME
The LED indicator will illuminate when the Babble is in broadcast mode. You can adjust the volume by turning the control. Your chosen volume setting will be confirmed vocally.

info

INFORMATION FUNCTION
Turn the volume control to 'info' and Energy Babble will tell you about itself, including a brief description of the different voices and types of content.

info

SLEEP MODE
To put the system into sleep mode turn the volume control to the '0' position. Energy Babble uses 4w when it is switched on, less than a clock radio. If you wish to turn Energy Babble off completely, simply unplug the power cable from the rear of the enclosure.

9

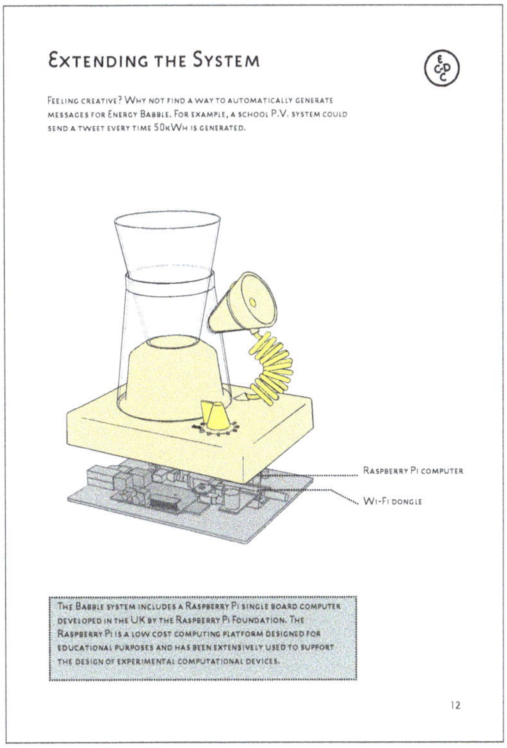

EXTENDING THE SYSTEM

Feeling creative? Why not find a way to automatically generate messages for Energy Babble. For example, a school P.V. system could send a tweet every time 50kWh is generated.

Raspberry Pi computer

Wi-Fi dongle

The Babble system includes a Raspberry Pi single board computer developed in the UK by the Raspberry Pi Foundation. The Raspberry Pi is a low cost computing platform designed for educational purposes and has been extensively used to support the design of experimental computational devices.

12

**Pages from the user guide that we
provided to all of the volunteers**

their experiences with the Babbles. In the background, the Babbles streamed their odd

...... The UK energy demand has gone up by 415 mega watts ...... ♩♫♪♫ ......
Tenesol Solar Panel 190W in South Africa ...... ♩♫♭♪ ...... Thanks for bringing
that up ...... ♪♫♩♫ ...... Joe from says. EU and UK reason our energy cost
twice that of the US? environmental cost, green taxes! price rise on all fuel bills,
transport of goods food! ...... ♩♫♪♪ ...... Tanya Ha from Melbourne says.
This Ron Tandberg cartoon kind of sums up Oz blindness to renewable energy
opportunities ...... ♩♫♫♩ ...... PARISFrench energy engineering firm Areva
SA and Spain's Gamesa Corporacin Tecnolgica SA Monday said they were joining
forces to create an offshore wind power business, embracing the consolidation
wave of an industry plagued by massive costs. The two groups will have an equal
share of the joint venture which still has no name ...... ♩♫♪♫ ...... Martin G.
Kamau from Nairobi, Mombasa, Kenya says. Progress is being made, though
painstakingly slow. The shift to geothermal, wind and LNG power plants is real!
...... ♩♫♭♪ ...... It is uncharacteristically warm this October Sid Valley. Hello
testing one, two, three do you hear me? ...... ♩♫♫♩ ...... US Army colonel:
world is sleepwalking to a global energy crisis A conference sponsored by a U S
military official convened experts in Washington DC and London warning that
continued dependence on fossil fuels puts the world at risk of an unprecedented
energy crunch that could inflame financial crisis and exacerbate dangerous
climate change. The Transatlantic Energy Security Dialogue, which took place on
10th December last year, was co-organised by a U S Army official, Lieutenant
Colonel Daniel L. Davis, operating in a private capacity, in association with former
petroleum geologist Jeremy Leggett, covener of the UK Industry Taskforce on
Peak Oil and Energy Security ...... ♩♫♩♪ ...... Recent message: I have an
energy babble and very lovely it is too. ...... ♪♫♩♫ ...... Good afternoon from
Derek at Sid Valley Energy Action Group. Nice feedback ...... ♩♫♪♫ ...... The
UK energy demand has gone up by 1036 mega watts ......

mix of voices and music as we tested them out. These odd devices were the culmination of

months of work in our studio — work oriented not so much towards discovering truths as

Studios, problems, publics

Alex Wilkie

What happens when the 'problem' of climate change and energy-demand reduction (as much a problem as it is a solution) enters and passes through a design studio? To further complicate, or nuance, the question: what happens when existing local community-based solutions of reducing greenhouse gas emissions in the UK by at least 80 per cent by 2050 are themselves re-problematised and re-determined as they pass through a university-based design research studio? Let me again rephrase the question: what happens when the problem-solving efforts and initiatives of UK energy communities form the basis of a 'multidisciplinary' funding initiative, led by Research Councils UK, to further support local community engagement with carbon reduction and environmental targets, which is then passed – as a research 'problem' – through studio-based practices involving an interdisciplinary team of design researchers and STS scholars? In this chapter, I explore how the Interaction Research Studio, as a centre of expertise in design research, engaged with the problematics of energy-demand reduction. In so doing, I aim to contribute to our understanding of how (design) studios can elicit climate-change publics.

The questions I pose above speak to any number of contemporary concerns about the nature of energy problems, knowledge production, and, not least, the nature of design research studios as the settings of epistemic practices: the logics, modes, and forms of accountability enacted as part of interdisciplinary engagements between design and science & technology studies (STS); the relationship between the interdisciplinary calls to upstream problem-formulation-solution dialogues (e.g. Rogers-Hayden and Pidgeon 2007); and the participatory role (engagement, inclusion, involvement etc.) of citizens and publics in deliberative and democratic knowledge practices. The questions also invoke disciplinary preoccupations such as whether design is best

characterised as a problem-solving or problem-setting discipline (e.g. Schön 1985; Wilkie and Michael 2015); how to acknowledge and understand the role of the studio as part of epistemic practices; and, in the case of sociology and STS (if, indeed, the latter can be called a discipline), what is the nature of the 'social' (Savransky forthcoming), or what counts as the empirical (Wilkie et al. 2015), and how to go about accessing both with methods or techniques that are, in part, constitutive of the 'object' of study? For the ECDC project, such preoccupations, as this book shows, were set alongside more mundane and practical challenges about how studio members with diverse skills and interests work together (which might be understood as 'distribution of labour' rather than 'division of labour'); where and how to bring together STS analysis with the making of research devices; and how to turn, or combine, reflections on invention into inventive practices? These questions also point to the empirical nature, conditions, and thick multiplicity of energy-demand reduction problems: something the project team encountered from the very beginning of the project through engagements with the local energy communities. Each community came with its own problem situation wrought by the event of climate change, such as inner-city immigration, energy poverty and literacy; the formation of formal organisational arrangements to collectively manage government funding; and the technical expertise required to install and manage solar PV cells, ground-source heat pumps, and wind turbines; as well as the practical problem of how to maintain the together-ness and composition of a 'community' as members' preoccupations and priorities alter, or members simply move elsewhere; or how composition – and the very nature of 'membership' – is reconfigured by the ongoing inclusion/exclusion of non-humans, e.g. infrastructure, technologies etc., and practices.

For readers familiar with the practices of STS, the typical way to get to grips with the questions and concerns that I've raised above is to work through some empirical examples and draw out a number of nuanced analytic points (around, for example, the interconnection between studios, problems, and publics). Not wanting to break with tradition, I intend to use the same approach here. This will involve a shift from recovering the studio as a substantive analytic topic for social and cultural research (Farías and Wilkie 2015b) to the view of the studio as a centre for the doing of design and social science, as well as a practico-theoretical reflection on the nature of 'problems as process', and how the design studio operates to resource the materialisation and mediation of publics. While I am wary of the now all-too-common studio-laboratory designation (a notable feature of university-based design research initiatives which arguably finds its apotheosis in pedagogic laboratory set-ups where designers learn to undertake the practices of synthetic biology), which is discussed in the aforementioned book (see chapters: Farías and Wilkie 2015a, Born and Wilkie 2015), drawing on the STS tradition of laboratory studies, with some caution and provisos, can provide a useful entry point into and contrast with knowledge and material practices enacted in design studios. Here, in particular, the view that an 'experiment is an event' (Latour 1999: 126) in which all the entities involved in laboratory processes are (partially) transformed and transform one another and, as an upshot, acquire their competencies, may yield insight into research events (Michael 2012a, Michael 2012b) that might take effect in, and transform, the design studio. This, then, is to take studios, problems, and publics as (epistemic, ontological, and aesthetic) processes – the parameters of which undergo transformation during interdisciplinary design/STS engagements. Perhaps one key way, as previously signalled, in which design studios diverge from modern scientific laboratories (for further contrasts see: Farías and Wilkie 2015a, Wilkie and Michael 2015) is that they entail further confusions between epistemological and ontological questions with explicit preoccupations concerning the aesthetic quality of studio practices and outcomes. In addition, if the constraints of the scientific laboratory and the design studio and what counts as an 'experiment' are divergent, i.e. if the 'event' of the scientific experiment is to raise the possibility of the claim 'Nature has spoken', then what is the corollary for design studios?

It follows, then, that if laboratories, as sites predisposed to making nature speak, play a key role in the articulation of new and enhanced publics as an upshot of the construction of facts and the introduction of novel entities and practices such as fermentation and vaccination, neurobiology, neuroendocrinology, particle physics, radio astronomy and so on, that studios, as another order of centres of expertise, also play a role in the expression of, what might be characterised as, aesthetic publics. I use aesthetics here, however, in a post-Kantian sense (Shaviro 2009) as non-cognitive ordinary experience (and thus attributable to all entities and phenomena) rather than as a question of reflexive human judgements (about taste or nature) – an approach which strikes me as particularly germane to the problem(s) of energy, which is much more than simply a matter of economic rationalization, as sociocultural approaches to energy consumption show. Seen from this perspective, studios can be grasped as heterogeneous and machinic processes that play a role in: the elicitation of publics accustomed to high modernist music (Born 1995); the stimulation of middle-class Kenyan civil society through post-ethnic 'world music' (Born and Wilkie 2015); or attempts, by way of the material production, broadcast, and reception of sound to achieve 'Reithian' auditory publics (Oswell 2008); or, in the more explicit case of design studios, the practical and material reconstruction of West German democratic society by way of industrial and graphic design (Spitz 2002); or attempts to manage obesity in populations through wearable technology (Wilkie 2014). Understood as such, studios can be received into the inventory of centres of expertise that are, according to some, the wellsprings of contemporary power and politics (e.g. Latour 1983: 168). Given the above, the role of the Interaction Research Studio in the ECDC project served as a centre of expertise that was brought to bear on the problem of energy-demand reduction, climate change, and its attendant energy publics in the making.

To work through the analytic formula 'problem-studio-public' that I have sketched out above, I focus on three vignettes drawn from my studio experience on the project. First, I show how one of the primary UK solutions to domestic energy-demand reduction, the 'smart monitor', was brought into the studio and found to be replete with practical complications and difficulties as a device charged with eliciting more responsible technical-aesthetic publics. Second, I discuss how the writing of the Energy Babble brief, approximately a third of the way into the project, worked to reconfigure the constraints of the energy-demand reduction problematic in the form of a generative lure for the production of technical-aesthetic responses – that would actively guide and shape the design of the Babble research device. Lastly, I reflect on how the injection-moulded enclosure of the Babble underwent an online and automated 'Mouldability Advisory' checking process to anticipate and iron out manufacturing problems and adapt to their tooling process. In doing so, I discuss how the resolution of technical-aesthetic details mediated the micro-physics of aesthetic publics.

Monitors
One of the first energy-demand reduction problems to enter the Studio came with an ostensible and grandiose solution. Early on in the project, after having collated and reviewed much of the scholarly and policy literature on the subject, we encountered the UK rollout of 'smart meters' to all UK households by 2020 as one of the key policy instruments

towards instrumental approaches to energy-demand reduction such as smart meters, and

(e.g. DECC 2009) with which the UK government intended to address carbon reduction. The reasoning underpinning the rollout of smart meters is predicated on a behavioural model of energy consumption where the 'energy user' responds positively to information feedback about their own energy use. In other words, when householders are confronted with real-time information about their energy usage they will respond in a rational and calculative way to reduce their overall consumption, such as identifying and reducing their use of particularly inefficient or demanding appliances such as kettles. According to studies of smart monitors and informational nudges, it was becoming apparent that the efficacy of smart monitors was fraught with problems (see for example: Abrahamse et al. 2005, Hargreaves et al. 2010b, Buchanan et al. 2015) – and notably the 'boomerang' effect (Schultz et al. 2007) whereby energy efficiency actually engenders increased consumption (e.g. leaving energy-saving lights on longer) – as well as the view that the integration of such technologies into households ignores the complex social, cultural, and technical settings and practices through which energy is supplied and consumed (Strengers 2008, Shove 2010, Hargreaves et al. 2010a, Shove and Walker 2014). Arguably, then, smart monitors fail to adequately materialise the problem of public engagement and participation in carbon reduction measures on two interrelated counts. First, by reducing the modalities of engagement and connection with climate change (Marres 2011) through the normative figuring of publics as comprising rational and calculative individual-as-energy-consumer constituents, and thus bracketing out other modalities of engagement with energy. Second, if the effect of smart monitors is to elicit such energy publics, then in application and end-use, smart monitors also fail to live up to the behavioural promise of achieving such constituents by not delivering on participation (as abstinence) made easy as part of routine consumption, for example owing to boomerang effects, and installation hindrances, as well as failings in efficacy and intelligibility of feedback.

Not satisfied with the aforementioned critiques of behavioural models of change and the sociality of monitoring technologies, not least because of the lack of alternative proposals, we set about exploring the complexities of smart monitors in use by purchasing an indicative set of monitors, installing them, and then using them for a relatively short period of time. Here, we tested a variety of monitors to reflect current market offerings including rudimentary models by Current Cost as well as the Wattson, which glowed different colours to display high or low consumption. This was followed by a short workshop, drawing on the notion of the 'script' (Akrich 1992), and held in the Studio, to visually trace and identify the 'distribution of competencies' (ibid. 207) required for the monitors to work as intended. Needless to say, the fact that Akrich's explication of technological scripts was, in part, advanced through a case study describing the attempt by the Ivory Coast government to achieve a governable

and spatialised public (by mobilising citizens) through the introduction and distribution of electricity monitors, was not lost on us.

The script analysis workshop revealed many issues and challenges faced through the installation and use of the smart monitors. In any number of ways and in virtually all cases, we failed to use the devices, whether through difficulties in setting them up in rental properties (see Liliana's 'My energy monitor: Chronicle of a failed attempt', this volume) with difficult-to-access and shared electricity meters, through to the inability to meaningfully equate the smart monitor readings, usage figures, or ambient colour, as in the case of the Wattson, with actual consumption. A particularly acute example of the latter being how the smart monitors failed to incorporate actual charges in read-outs, owing to the complexity and variability of energy billing, which placed the burden of calculating real-time cost benefits on the end-user. The practical question of how to intervene in household carbon reduction via the mobilisation of publics engaging with climate change in community settings therefore revealed itself, to echo Akrich, as productive of a redistribution of both technical and aesthetic problems, aligned with technical and aesthetic publics.

Brief
The Energy-Babble brief, presented in the 'Designing' section of this book, marked a key milestone in the ECDC project. The brief emerged after having reviewed and explored literature, technical and technological conditions, available solutions, related and implicated designs, community members' energy-related practices and expectations, as well as any other material the team felt relevant or informative – not excluding the fanciful or whimsical. Much of this investigation was brought together in the form of a series of design workbooks that variously included observations, insights, and analysis, as well as notional designs visualised through collage, sketches, annotations, and so forth. The brief itself was collectively written and overtly succinct: 'Design an Energy Babble system that displays material, collected from some combination of individual, community and public sources to open and promote constructive affect and involvement in energy-reduction issues and orientations. More specifically, the system should support an understanding of and practices related to energy-demand reduction.' At the same time, it included a number of footnotes (see page 49) that sought to clarify or elaborate on particular notions and terms, and introduce different viewpoints.

Clearly, the brief produced by the team was atypical and diverged from convention (see Dorst and Cross 2001). Arguably, design briefs typically act to make visible and accountable the interests, obligations, and requirements of both the client and the appointed designer/design team, setting out the terms of reference of the work to be done and defining what is deemed acceptable. Taken as such, design

instead sought simultaneously to draw together and share information among 'energy commu-

briefs can act as the inscription of contractual obligation to a problem, or 'problem space' (ibid. 2001: 434). Although university-based design studios can be likened to 'inscription devices' (Wilkie and Michael 2015: 30) in that they harbour and are made possible by various modes of writing practices, as well as the production of scholarly publications, the Energy-Babble brief suggests that briefs can operate very differently. The brief, mentioned above, demands a different set of obligations: obligations that are not formatted by the requirement to adequately respond to the demands or interests of a client. Of course, the obligations of the ECDC project were formulated in the grant application – which can be read as, or likened to, a brief.

Here, a certain (mis)reading of Whitehead's metaphysical notion of 'proposition' provides a useful way to understand the role of the ECDC brief. For Whitehead, propositions are a mixture of the actual and the virtual, a 'hybrid between pure potentialities and actualities' (Whitehead 1978 [1929]: 186-186), the prime function of which is to act as a 'lure for feeling' (ibid. 25), which is proposed to a process (as problem), such as energy-demand reduction. Propositions therefore play a particular part in the constitution of what Whitehead calls 'actual entities' – which may find completion, or satisfaction in, for example, a particular design or designed outcome. Whitehead also states that it 'is more important that a proposition be interesting than that it be true' (1967: 244). Understood in a propositional manner, the ECDC brief can be seen as a lure for carbon reduction interests and demands: it is suggestive of how a design might come about and what is required of it, though with some degree of uncertainty. In other words, the brief concretely acts in the process of inventing and defining problems out of a complex topology of concerns (see Schön 1991: 40) all related to the question of community engagement in energy-demand reduction. Taken one step further, and paraphrasing Whitehead, the brief can be further understood as a practical lure for problems (problems, here, defined themselves as lures or demands) where the actual conditions of public and community engagement with carbon reduction are made relevant through the practices of the designers and the obligation to bring into being a research device.

The Babble brief, however, added another layer of complexity to the articulation of community (and public) engagement with carbon and energy-demand reduction. Rather than act as a lure for the emergence of a well-defined problem-solution coupling (which, in the case of smart monitors, can be manifested as problematising), the brief sought to engender the materialisation of a research device. The aim of the device was to problematise and entertain the possibility of novel relationships that community members have with climate change, and, further, to promote alternative engagements and understandings by way of some combination of scraped web content (sourced from different platforms and from a variety of implicated actors)

and community input. The brief, therefore, contained aspects of cosmopolitical (Stengers 2011) concerns, whereby, rather than close down the limits or framings of carbon-reduction problems by encouraging a well-defined solution (which is, unsurprisingly, a feature of policy discourse and instruments) by encouraging a well-defined solution, the brief sought to promote the design of a speculative device that would give voice to disparate energy and environmental problems, interests, practices, realities, and publics that may coexist or be in the process of coming into being. Put in more practical terms, the cosmopolitical proposition (rather than the problem) posed by the ECDC brief concerned the actual and the possible composition of energy communities and publics.

Enclosure

The upshot of the brief came in the form of the Energy Babble device, which emerged over the course of a year by way of numerous design-related activities, including, but not limited to: interaction, graphic, 'product', and sound design as well as hardware (component) and software design. The manufacturing of the enclosure of the Energy Babble provides another example of how the problems associated with energy-demand reduction get rendered through the aesthetic and technical qualities of the casing. As is variously shown throughout this book, the aesthetic quality of the Energy Babble did not conform to the normative role of design to materialise ease-of-use and aesthetic 'harmony' as is the case with other mundane domestic appliances and devices. Rather, the aesthetic qualities of the device, alongside its technical functionality, acted to question the possibility of other modes of engagement with energy-demand reduction and to problematise existing solutions as 'involvement made easy' (Marres 2011): e.g. that involvement made easy requires particular achievements during end-use, and that its efficacy is far from easy.

The enclosure for the Energy Babble had to meet a certain set of requirements, such as housing a Raspberry Pi computer, a USB sound card, a Wi-Fi dongle, and a USB loudspeaker, as well as provide input/output for data, wired and wireless communications, and power as well as interactional elements for 'users', such as an 'on' switch, volume control, and a microphone handset. Added to these requirements were the cost and time implications of batch-producing more than thirty units of the device. Although the team set out with the expectation that the enclosure could be fabricated using in-house 3D printers, it soon became apparent that outsourcing the manufacturing of the enclosure would be more be efficient and economical whilst simultaneously addressing issues the Studio experienced on previous devices around how 3D-printed components somewhat rapidly deform whilst perishing. This decision also provided the opportunity to work with the visual, material and tactile qualities of injection-moulded parts, which had recently been made available by the company Protomould UK for prototyping purposes (short

nities' while problematising energy, communities, and even the idea of sharing information

batch production runs of parts). Protomould's novel services for automating and reducing the cost of injection-moulding tooling opened the possibility of using manufacturing techniques that are typically only economically feasible for large-scale production for short-run batches. The decision to work with Protomould, however, brought about its own challenges related to developing the expertise necessary to specify the design of an injection-moulded part, and achieve compatibility with the company's fully automated online mouldability analysis and advisory system.

The decision, then, to proceed with injection moulding brought about additional processes during the instauration (Souriau 2015: 128-129) of the Babble – processes that required members of the Studio, notably Liliana, to acquire the skills and expertise to successfully mediate, translate, and transform the problems of aesthetic qualities into technical engineering solutions. Processes that would necessarily involve changes, however minor, to the form of the device. This involved, in part, attaining a new set of competences with 3D modelling software in order to meet the variety of demands placed on the enclosure. The injection-moulding demands included meeting the required draft angles, tolerances, changes in wall thickness, size of bosses, rebates, and gussets, and ensuring correct moulding flow so as to avoid warping and shrinkage. It would also allow for textured finishes as well as facilitating the reliable removal of the component from the moulding tools and minimising injection stresses. This process also involved specifying the best possible 'mating' details between the enclosure and the components that it would house so that parts had reliable internal structural support and registration (e.g. bosses that could support the metal insert threads for self-tapping screws required for affixing components to the enclosure).

After a visit to Protomould by Studio members to review manufacturing facilities and gather advice about certain details of the enclosure, such as apertures for cabling and input/output (I/O) ports, a two-month process of mouldability analysis and resolution ensued. 3D CAD files were uploaded to the fully automated 'Protoquote' service, which analysed them for compatibility with the injection-moulding process and returned advisory reports which used annotations to illustrate any issues identified. The reports also included updated costings based on the submitted design. The illustrations contained in the report featured pull-outs and coloured indicators that highlighted specific aspects of the design that required adjustment to meet the demands of moulding, as well as how the geometry of the CAD model would differ from the final product due to milling processes. After multiple uploads, reports, and adjustments, the enclosure was finally ready for manufacturing, and in less than two weeks the Studio took delivery of the parts. According to Liliana, who had never designed an enclosure for computational components before, the Protoquote process was akin to an online CAD training course (see also Andy's account of the microphone

casing design, this volume) with no human involvement or response on the part of Protomould. Each time she received a new advisory report she found herself having to apprehend new technical vocabulary (e.g. 'draft angles', 'sink marks', 'voids', 'ribs', 'gussets', 'drag marks' etc.), appreciate the moulding specifications, and then problem-solve meeting the specification by adjusting and redesigning aspects of the enclosure. For example, the bosses that would hold the thread insert for the self-tapping screws and provide mating support for the components proved to be too tall and they lacked enough angle to allow for the flow of the thermoplastic. In order to correctly specify the bosses, however, Liliana had to source and consult various resources so as to negotiate and coordinate the implications that redesigning the bosses would have on the rest of the enclosure without occasioning too many changes to the overall design. Although Protoquote indicated problems, it did not provide guidance on how to overcome the problems.

The case of the enclosure provides yet another way to grasp the relations between studios, problems, and publics. The notion of instauration which I mentioned previously, is useful here as it underscores the way in which the aesthetic 'design' of the enclosure emerged through a set of inter-relations between a designer, a 'design', CAD software, the requirements of components, the injection-moulding process, the attributes of thermoplastic, and the automated review system – each of which is active in determining the final nature of the enclosure. To instaure the enclosure is not a hylomorphic process in which a given design is merely realised in form (Sánchez-Criado et al. 2014: 7). Rather, the enclosure comes into being through the trials of the review system, the capacity of components to determine their requirements, the synthetic attributes of thermoplastic, Liliana's acquisition of engineering skills, and the ability to negotiate aesthetic and technical solutions (as determination and mixture). In other words, this is a (somewhat mundane) relational semiotic-material process in which the various elements and practices are becoming with one another during the making of the enclosure and the resolution of aesthetic-technical problems. Echoing Sánchez et al. (ibid.) this account of design-in-practice shows how the 'felicity conditions' (Austin 1962) are made and become during the process of design, rather than being given. Furthermore, the instauration of the Energy Babble enclosure raises the possibility of engendering energy communities (and publics) through the aesthetic-technical particularities of the device. If problems (and publics) are posed by events (Savransky forthcoming), such as climate change, then the example of the Energy Babble enclosure shows the detailed work through which problems are concretised and publics shaped, contributing to and 'thickening' the problem.

Concluding Remarks

In conclusion I return to the questions posed regarding the problem of climate change and how it enters into and

– and all this in a set of batch-produced devices that were aesthetically and functionall

passes through the studio. In doing so I show how the problem gets variously enacted through the description of smart monitors, the formulation of a design brief as well as preparation for the manufacturing process of a research device, each of which adds specific aesthetic as well as technical problems, raising the possibility of aesthetic publics along the way. The analysis of the smart monitors revealed that, in use, these household devices, more often than not, failed to live up to the promise of aesthetic satisfaction of calculative energy use. Rather than accomplishing an easy-to-use solution to the problem of climate change, the smart monitors occasioned a new set of mundane problems associated with installation and usage. Rather than seek a neat problem-solution coupling and reduce the problem of climate change to easy-to-use demand reduction, the brief, which I characterise as a propositional lure for problems, operated to open up the problem-space of community as a cosmopolitical question concerning the composition and nature of energy communities and publics. Whereas smart monitors, arguably, simplify the problem to a technical question of user-device efficacy, the brief puts into play the possibility that a research device might give shape and voice to other entities and actors involved in climate change and energy-demand reduction. Studio-based interdisciplinary design research, here, shifts the register – or, indeed, logic – of involvement from an aesthetics of effortlessness (ease-of-use of a domestic appliance) where well-defined participants (user-device) behave to an awkward and abstruse aesthetics in which those involved are not entirely known if at all and their satisfaction (concrescence and efficacy), so to speak, and mode of behaviour is not guaranteed. Consequently, the detailed design specification of the enclosure explicates the work entailed in materialising the grounds of possibilities – or felicity conditions – through which speculative energy publics might be elicited aesthetically – publics which are heterogeneously nurtured and heterogeneously composed.

It follows, then, that if the event of the scientific experiment is to enable the claim 'Nature has spoken' to be made, the event of interdisciplinary design research raises the possibility of the declaration and assertion of new (human and non-human) members of (new) energy publics and new energy-related practices. As such, the design studio itself can be understood as a device for recomposing problems, problem situations, and possible solutions.

Acknowledgements
I would like to thank Mike Michael and Martin Savransky for their insightful comments on various theoretical aspects of this essay. I would also like to thank Liliana Ovalle for generously sharing her experiences of the Protoquote service.

References
Abrahamse, W., et al., 'A Review of Intervention Studies Aimed at Household Energy Conservation', *Journal of Environmental Psychology*, 25(3) (2005): 273-291.

Akrich, M., 'The De-scription of Technical Objects', in W. Bijker and J. Law, eds, *Shaping Technology/Building Society: Studies in Sociotechnical Change* (Cambridge, MA: MIT Press, 1992) pp. 205-24.

Austin, J. L., *How to Do Things With Words*, 2nd edn (Oxford: Clarendon Press, 1962).

Born, G., *Rationalizing Culture: IRCAM, Boulez, and the Institutionalization of the Musical Avant-Garde* (Berkeley: University of California Press, 1995).

— *Uncertain Vision: Birt, Dyke and the Reinvention of the BBC*, New edn (London: Vintage, 2005).

— and A. Wilkie, 'Temporalities, Aesthetics and the Studio: An Interview with Georgina Born', in I. Farías, and A. Wilkie, eds, *Studio Studies: Operations, Topologies and Displacements* (Abingdon, Oxon; New York: Routledge, 2015) pp. 139-55.

Buchanan, K., R. Russo, and B. Anderson, 'The Question of Energy Reduction: The Problem(s) with Feedback', *Energy Policy*, 77(0) (2015): 89-96.

DECC, *Smarter Grids: The Opportunity* (London: Department of Energy and Climate Change, 2009).

Dorst, K., and N. Cross, 'Creativity in the Design Process: Co-Evolution of Problem–Solution', *Design Studies*, 22(5) (2001): 425-437.

Farías, I., and A. Wilkie, 'Studio Studies: Notes for a Research Programme', in I. Farías, and A. Wilkie, eds, *Studio Studies: Operations, Topologies and Displacements* (Abingdon, Oxon; New York: Routledge, 2015a): 1-21.

—, eds, *Studio Studies: Operations, Topologies and Displacements*, (Abingdon, Oxon; New York: Routledge, 2015b).

Hargreaves, T., M. Nye, and J. Burgess, 'Making Energy Visible: A Qualitative Field Study of How Householders Interact with Feedback from Smart Energy Monitors', *Energy Policy*, 38(10) (2010a): 6111-19.

— 'Understanding How Householders Interact with Feedback From Smart Energy Monitors – Opening the Black Box of the Household', Paper Presented at *Cultural Economies of Energy Consumption*, University of Manchester, 2010b.

'finished'. The Babbles in a way already had multiple identities: as embodiments of what

Latour, B., 'Give Me a Laboratory and I Will Raise the World', in K. D. Knorr Cetina, and M. Mulkay, eds, *Science Observed* (Beverly Hills: Sage), 141-70.

— *Pandora's Hope: An Essay on the Reality of Science Studies* (Cambridge, MA: Harvard University Press, 1999).

Marres, N., 'The Cost of Public Involvement: Everyday Devices of Carbon Accounting and the Materialization of Participation', *Economy and Society*, 40(4) (2011): 510-33.

Michael, M., 'De-signing the Object of Sociology: Toward an "Idiotic Methodology"', *The Sociological Review*, 60 (2012a): 166-83.

— '"What Are We Busy Doing?" Engaging the Idiot', *Science, Technology & Human Values*, 37(5) (2012b): 528-54.

Oswell, D., 'Concrete Publics? Noise, Phantoms and Architectures of Radio and Television Reception from the 1920s to 1960s in the UK', Paper Presented at *The Physique of the Public*, Goldsmiths, University of London, 2008.

Rogers-Hayden, T., and N. Pidgeon, 'Moving Engagement "Upstream"? Nanotechnologies and the Royal Society and Royal Academy of Engineering's Inquiry', *Public Understanding of Science*, 16(3) (2007): 345-64.

Sánchez-Criado, T., D. López, C. Roberts, and M. Domènech. 'Installing telecare, installing users: Felicity conditions for the instauration of usership.' *Science, Technology, & Human Values*, 39, no. 5 (2014): 694-719.

Savransky, M., 'The Social and Its problems: On Problematic Sociology', in N. Marres, M. Guggenheim, and A. Wilkie, eds, *Inventing the Social* (Manchester: Mattering Press, forthcoming).

Schön, D. A., *The Design Studio: An Exploration of its Traditions and Potential* (London: RIBA Publications Limited, 1985).

— *The Reflective Practitioner: How Professionals Think in Action*, 2nd edn (London: Ashgate Publishing Limited, 1991).

Schultz, P. W., et al., 'The Constructive, Destructive, and Reconstructive Power of Social Norms', *Psychological Science*, 18(5) (2007): 429-34.

Shaviro, S., *Without Criteria: Kant, Whitehead, Deleuze, and Aesthetics* (Cambridge, MA: MIT Press, 2009).

Shove, E., 'Beyond the ABC: Climate Change Policy and Theories of Social Change', *Environment and Planning A*, 42(6) (2010): 1273-85.

Shove, E., and G. Walker, 'What is Energy For? Social Practice and Energy Demand', *Theory, Culture & Society*, 31(5) (2014): 41-58.

Souriau, É., *The Different Modes of Existence* (Minneapolis, MN: Univocal, 2015).

Spitz, R., *HfG Ulm: The View Behind the Foreground: The Political History of the Ulm School of Design*, 1953-1968 (Stuttgart; London: Edition Axel Menges, 2002).

Stengers, I., *Cosmopolitics 2* (Minneapolis; London: University of Minnesota Press, 2011).

Strengers, Y., 'Smart Metering Demand Management Programs: Challenging the Comfort and Cleanliness Habitus of Households', Proceedings of the *20th Australasian Conference on Computer-Human Interaction: Designing for Habitus and Habitat*, Cairns, 2008.

Whitehead, A. N., *Adventures of Ideas* (New York; London; Toronto; Sydney; Singapore: The Free Press, 1967).

— *Process and Reality: An Essay in Cosmology*. Gifford Lectures of 1927-8; corrected edn (New York: The Free Press, 1978 [1929]).

Wilkie, A., 'Prototyping as Event: Designing the Future of Obesity', *Journal of Cultural Economy*, 7(4) (2014): 476-92.

Wilkie, A., and M. Michael, 'The Design Studio as a Centre of Synthesis', in I. Farías, and A. Wilkie, eds, *Studio Studies: Operations, Topologies, Displacements* (Abingdon, Oxon; New York: Routledge, 2015), pp. 25-39.

Wilkie, A., M. Michael, and M. Plummer-Fernandez, 'Speculative Method and Twitter: Bots, Energy and Three Conceptual Characters', *Sociological Review*, 63(1) (2015): 79–101.

designed artefacts we hoped people might find enjoyable and intriguing. Meanwhile, we had started to contact our colleagues in the energy communities to tell them that the device

CIRCULATING

We delivered the packaged Babbles to our partners. These moments of unboxing were at times taken by our communities as opportunities for social gatherings – in one case we were guests at a dinner party – while other times they were occasions for more formal group meetings. Gradually the devices became distributed across the communities: on a table in the school library, amongst the books in a bedroom, in the window of a community information shop.

Deployments were not always straightforward. On more than one occasion we wrangled with wireless networks, hardware at times failed, and sometimes units were turned off with the hoover. One by one, however, the devices started talking.

Responses to the Babbles came immediately as people unboxed the devices. Faced with these strange devices, they started to imagine what living with them might be like and how they might use the Babbles to further their causes. In the months that followed, occasional phone calls, emails, and site visits provided glimpses into peoples' experiences. This culminated in our visits to pick up the devices, as discussions of what people loved and hated about them flowed into much wider conversations of the issues of achieving an energy community.

Deploying

Having finally produced the Babble devices, we set about getting them into the homes and neighbourhoods of energy community members. For the most part, this meant coordinating visits with their ongoing activities and events, which was sometimes tricky. It took us several months to travel to all the communities, hand over the Babbles, assist with installing them, and join in discussions about them. Eventually we managed to deploy three or four Babbles to each community, where they were installed mainly in homes, but also two schools and a pub, with a few left behind to be distributed locally.

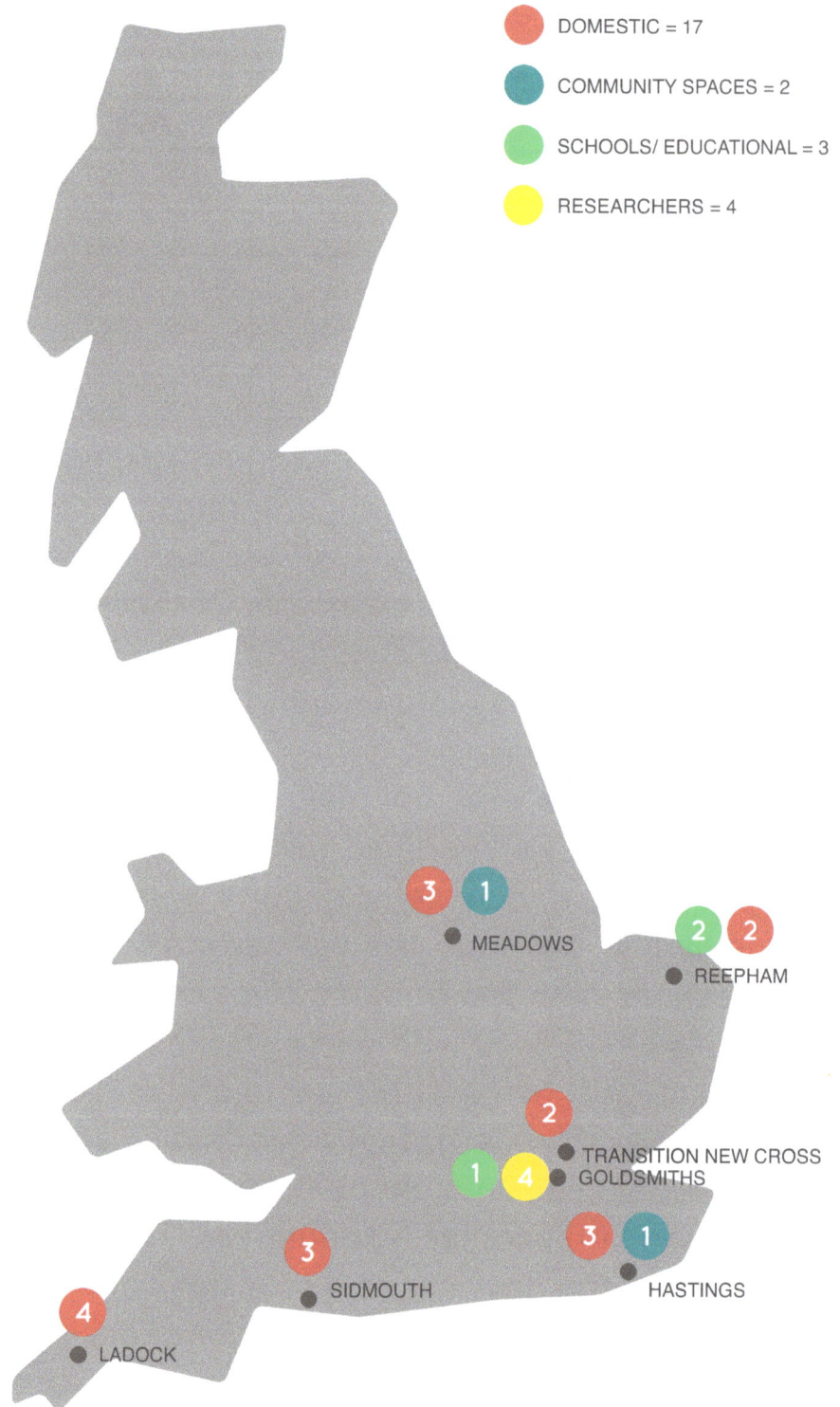

DOMESTIC = 17

COMMUNITY SPACES = 2

SCHOOLS/ EDUCATIONAL = 3

RESEARCHERS = 4

3 1
● MEADOWS

2 2
● REEPHAM

2
● TRANSITION NEW CROSS
1 4 ● GOLDSMITHS

3 1
● HASTINGS

3
● SIDMOUTH

4
● LADOCK

Map of the range of
deployment settings across
the energy communities.

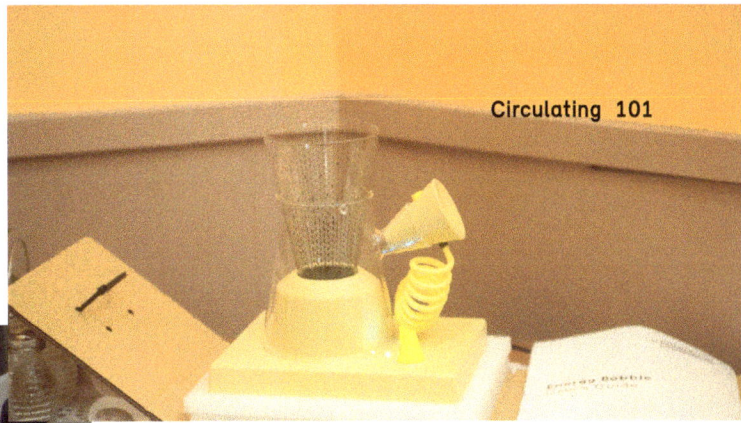

**Energy Babbles on the
move to their new homes**

ea or even dinner, and visited the various sites where the Babbles would live. Installing

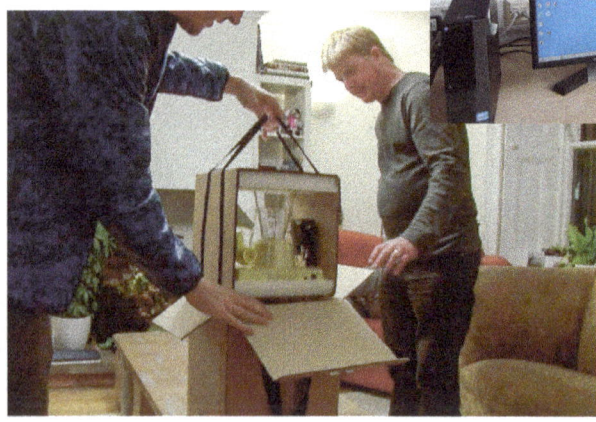

The devices were presented to groups
of volunteers in pubs, community
centres, and households.

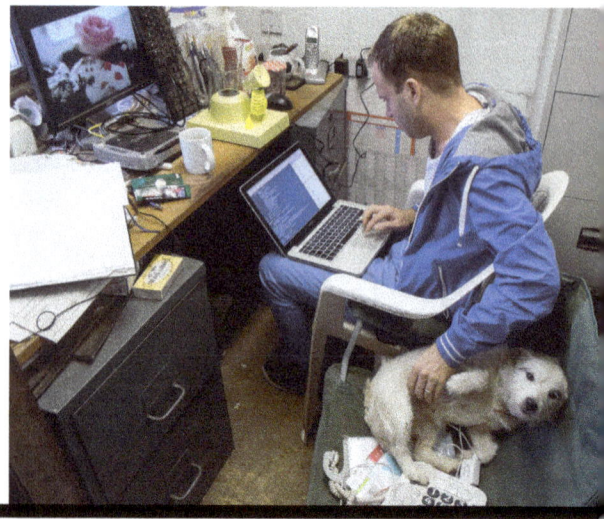

the Babbles often proved problematic. We needed to find power, connect to routers, bypas

'Engaging with' and 'engaged by': Publics and communities, design and sociology

Tobie Kerridge & Mike Michael

The ECDC project, as with other work that falls under the, albeit problematic, banner of speculative design, can be understood in terms of the broad interdisciplinary (and interlinked) fields of 'public understanding of science' (PUS) and 'public engagement with science and technology' (PEST). The members of the energy communities with which we collaborated can be recast as members of 'the public' – they could, after all, be juxtaposed with the government and industry energy experts who generated the technologies and knowledges (not least statistical and technical accounts) of energy-demand reduction. Our various interactions with the members of energy communities (e.g. initial meetings, probe workshops, site, and ethnographic visits, the introduction of the Babble) could be interpreted as attempts to access complex and critical public 'understandings' of, and engagements with, energy-demand reduction, not least in relation to the institutional settings through which energy-demand reduction is mediated (as a set of favoured practices, or competitive 'tendering' processes, for example).

However, in this essay, we also begin to unravel this picture by tracing some of the convolutions of these communities both 'internally' (they were 'stratified' publics with their own 'experts', 'advocates', and 'laypeople' and 'externally' (they were embroiled in contorted relations of competition and collaboration with other communities, and other constituencies). Our empirical argument is that while PUS and PEST can certainly serve to illuminate the social and material processes in which energy communities are involved, these observations remain limited. Once the broader environment is taken into account – that is, when these communities are situated within their wider networks or assemblages (including those of the present research project) – new and novel empirical insights emerge. Our theoretical argument is that this suggests that the conceptual parameters of PUS and PEST need to be re-thought in relation to both the present ECDC case study, and any given study.

In what follows, we begin with a brief outline of PUS and PEST, and how these map onto practice-based design research. In particular, we discuss how the Energy Babble might be said to 'work' in relation to the 'core' and emergent concerns in the fields of PUS and PEST. We then go on to open up the picture of the research, exploring several of the ways in which the ECDC project engaged with, and was engaged by, members of the energy communities. In the process, we encounter a much more variegated set of social and material relations which is usually neglected – one that is relegated to the hinterland of the research, as it were. We thereby begin to throw a very different light on the research itself, the energy communities, and PUS/PEST as viable framings.

PUS and PEST

'Public understanding of science' was initially oriented towards the study of lay people's grasp of scientific facts and procedures. Quantitative methods such as questionnaires were the main research tool used to measure levels of – or deficiencies in – 'scientific literacy'. One overarching concern was that without proper understanding of science, there would be less support for science (and scientific institutions). Partly in response to this 'deficit model', a critical or interpretative PUS developed in which the focus of analysis shifted to address the complex relations between science and society in general, but also particular publics and particular scientific institutions. With the use of such qualitative methods as interviews, focus groups, and ethnography, critical PUS began to excavate the tensions that arose between, on the one hand, the folk knowledge or lay expertise of publics, and, on the other, scientific institutions' oftentimes overzealous advocacy of scientific knowledge. Especially important was the tracing of how scientific institutions confidently insisted on the objectivity of their knowledge, often at the expense – that is, the derogation – of lay local knowledges. Not only did this threaten local identities, but it also led to a dilution of publics' trust in the credibility of scientific institutions, especially when their expert pronouncements turned out to be problematic (based as they were on scientific studies that were contingent on technical and social assumptions that did not always readily generalise). One implication was that scientific institutions needed to become better able to accommodate the knowledges and concerns of publics (for overviews of the evolution of PUS, see, for instance, Wynne 1995; Irwin and Michael 2003; Bucchi and Neresini 2008).

This accommodation in which the boundaries between science and society, and scientific institutions and publics, became, in one way or another, eroded has been discussed in various ways. For some scholars, there was a systemic change in which, for example, the increasingly socially embedded character of scientific and technological problems (for example, climate change, or nanotechnology) led to a greater role for lay or 'non-expert' actors in addressing these problems – what Nowotny et al. (2001) called 'Mode II Science'. At a more microsociological level, some authors suggested that there were already many examples in which lay and expert actors operated together. For example, some publics were sufficiently knowledgeable as to serve in the technical delineation of medical problems (Epstein 1996; Arksey 1998; Callon et al. 2001), while others suggested that scientific and technological controversies entailed antagonistic groupings – or assemblages – composed of a variety of publics and experts, including legal, economic, media, as well as scientific (Irwin and Michael 2003).

In addition, practitioners of critical PUS also became more proactive, advocating processes and procedures whereby the public's voice could be better integrated into expert deliberation and scientific policymaking. This is the era of 'public engagement with science and technology' (PEST), characterised by both the development and testing of a series of deliberative and participatory techniques for enabling public engagement including consensus conferences, citizens' juries, deliberative polling, card-based group discussion, etc. (Hagendijk and Irwin 2006; Chilvers 2008). However, running alongside this were a series of critiques which drew out a number of limitations with these methods, for example: their de facto lack of purchase within the decision-making process; the impoverished version of democracy they assumed; their proceduralism; their still overly narrow conceptualisation of the public (e.g. Lezaun and Soneryd 2007; Felt and Fochler 2010; Michael 2009; Marres 2012). This is still very much a live area (as evidenced by the recent volume edited by Chilvers and Kearnes 2016), and it is one in which design – in various guises – has had

an increasingly prominent role to play. We turn to this in the next section in which we also discuss how ECDC maps onto, and indeed can partially re-envision, PEST.

Design and PEST

Design is inherently concerned with 'publics' who are often translated into the terms of users – they use the designs developed by designers. However, the relationship between designers and users takes numerous forms (which are not always easy to disambiguate). Thus, at one pole (of a nominal dimension of designer-user interaction), the designer simply imagines what the user-public might want, need or desire (whether that be explicit or implicit). Here, the designer draws on their expertise and, to some lesser or greater degree, a model of the user, to design objects or services that they believe are best suited to particular functions. Obviously enough, the model of the user can vary widely, from a narrow consumer to a world citizen (e.g. Papanek 1984), with corollary political implications. At the other pole, designers directly engage with publics in order to design their artefacts or services. In the case of participatory design, or co-design, not only might prospective users contribute to how best to realise particular or specified goals or objectives, but they might also have a voice in redefining the very nature of those goals and objectives (e.g. Telier 2011; Storni et al. 2015). At something of a tangent to this engagement dimension is a set of design practices where the aim is not to provide an orthodoxly 'functional' object or service (however 'function' is delineated), but to develop entities and interactions which serve as provocations of some sort. Here, a particular object, for instance, might operate in ways that do not make obvious 'practical sense' – ways which might be obscure, or ambiguous, or playful, or confusing. The point is that through an 'engagement' with the object, the 'user' ideally comes to critically rethink, say, the direction of technological futures (e.g. Dunne 2005; Dunne and Raby 2013), or else opens up the meanings that attach to certain activities or phenomena such as 'advertisements', or the 'neighbourhood', or air traffic noise (Gaver et al. 2008; Michael and Gaver 2009). To be sure, while public engagement with the object can range from a brief encounter at a gallery all the way through to a sustained interaction in which users live for extended periods with the object, the general aim of opening up the possibility for otherwise unarticulated views remains.

ECDC falls within the latter of these design approaches – what is sometimes called 'speculative design' (though as we have stressed, this moniker is deeply contestable). Throughout this volume, we have documented how ECDC engages with publics (through probes, probe workshops, site visits, etc.) in order to design a prototype (the Energy Babble), which is, however, developed at a remove from those publics. The Babble emerged out of a series of dense and convoluted discussions that drew on the materials derived from our energy communities, but also with reference to various other sources (from design history, through the

and explain how they were intended to engage people with energy by opening questions rather

politics and policy of energy-demand reduction, to research into energy communities conducted elsewhere). The aim was to construct something that was sufficiently opaque, playful, and ambiguous so as to enable users to open up 'energy-demand reduction': insofar as the Energy Babble proved to be 'idiotic' (see 'Design and Science & Technology Studies' essay), it could facilitate the unravelling of the meanings of energy, community, information, communication, reduction, etc. In relation to PEST, ECDC certainly aims to give voice to the public but it is not a voice that is necessarily directly relevant to the standard policy issues that surround 'energy-demand reduction'. Rather, it is in keeping with the ethos of speculative design (e.g. Michael 2012), speculative methodologies more generally (Michael in press; Michael et al. 2015), and the ambition of speculative approaches to interfere with 'streams' of technology innovation (Kerridge, 2015). ECDC can be said to be concerned with facilitating the generation of more 'interesting questions' or 'inventive problems' that open up the very meanings and practices that attach to 'energy-demand reduction'.

Yet, this picture of ECDC as an example of a 'designerly' elaboration of PEST does a disservice both to the design-oriented articulation of the project's relation to the public, and to the 'publics' we engaged with – publics which, as complex and interrelated communities, were immersed in a nexus of conditions with and against which they were obliged to operate. From the design perspective, the communities were not necessarily 'publics' – that is, groupings that stood in contrast to, and engaged in some sort of struggle with, 'expertise'. And the Babble was not simply a 'tool' through which to 'engage' publics (however speculatively), but a design object which sought to embody and speak to a variety of 'design issues' (the use and integration of particular technologies, algorithms, platforms, etc.) as well as engage with 'social issues' (open access to technologies, fuel poverty, the uptake of renewables, etc.). With regard to 'publics', the above account is all too linear – nuanced relations were developed with particular members of the communities, and it was clear from the outset that the project was as much 'being engaged by' community spokespeople, as we were 'engaging with' those communities. The Babble could thus be seen not only as a 'research device', but also as a 'sociopolitical device' being deployed within and across communities as members went about the local processes of 'making' and 'remaking' energy communities.

In the following sections, we reflect on these entangled processes of 'engaging with' and being 'engaged by' the energy communities. This can certainly be thought of in terms of the 'research event' discussed in the 'Design and Science & Technology Studies' essay, and the mutual effects of researchers and practitioners. However, as we shall also see, this was not a straightforward process, not least insofar as that 'research event' emerged as something other than 'research'. By way of a preview, we can note:

that the relations between 'energy communities' and 'researchers' was a fraught one; that the relations between energy communities was challenging; that we were not engaging with singular communities, let alone publics; that the 'leaders' or 'spokespeople' of energy communities were engaging us in particular ways in part to enact a particular trajectory for 'their' community; and that we ourselves as 'researchers' did not always have a consistent or coherent view of the communities or, indeed, of the project.

'Energy communities' and 'researchers'

From the outset, and through the formulation of its case for support to RCUK, the ECDC project sought to recruit and treat as respondents a subset of the 'communities' engaged in energy-demand reduction who had received support from DECC as part of its Low Carbon Communities Challenge (LCCC) competition. A total of twenty-two groups had received DECC grants for demand-reduction measures. We (as researchers) were introduced to, and supported in developing relationships with, the LCCC competition winners (the energy communities). However, it soon became apparent that any such relationships required sensitivity and careful negotiation. Here, we draw on events facilitated by the research councils, DECC and UKERC, to consider the complex nature of relations between the energy communities and their researchers.

The Energy and Communities Research Workshop was an initial meeting in November 2011, and perceived by us as something of a matchmaking opportunity for research groups and communities. Following project 'pitches' by research groups, a set of prearranged, themed workshops supported various aspects of collaboration, including one titled 'Recruitment, Recognition, Valuing People'. Here, conversation was substantively led by community members, and covered a range of topics, including what was viewed as the problematic distinction between the researchers and the communities (which was seen to be patronising, or undervaluing local insights), the reductive nature of the term 'communities' (which simplified local coalitions that might include end-users, entrepreneurs, local government representatives, etc.), the 'plagiarist' nature of academic research (where narratives and practices are taken from communities as findings by researchers), and where academic research projects were seen to be an inappropriate use of resources (not supporting communities' core requirements, conducting ill-conceived or ineffective activities, etc).

A second meeting, convened by UKERC in Oxford in October 2011, sought to operationalise the misgivings and reticence of energy communities (now reconceptualised – perhaps more constructively and actively – as 'practitioners'). The meeting entailed an orientation document around the issue of 'Engaging Practitioners', which was sent to participating researchers in advance of the meeting, and which drew upon a survey that elicited practitioners' views

than suggesting answers. This was a tricky prospect, because some of the community members

on being researched. Issues for consideration in the framing of activities at the meeting included: the need for clarity regarding the 'benefits to communities of engaging with researchers'; a proposal that 'practicable, actionable steps can be recommended' to practitioners by researchers; and the observation that academics 'appear reluctant to fund legitimate and valuable input from practitioners'. Indeed, regarding the fiscal concerns of the last point, and against the backdrop of government's failing enthusiasm for renewables at that time (along with the tailing off of funding from DECC and others), the academics' projects seemed to be bountifully resourced at a time when practitioners' resources were becoming stretched and their activities precarious.

Additionally, in dealing with one of the most researched groups in the UK, we can interpret this lack of enthusiasm (and at times antipathy) from energy community practitioners, at least partly as an indication of research fatigue (Clarke 2008). Despite these issues (and others addressed in this volume), our abiding sense is that ECDC managed to sustain the engagement, or at least the polite acquiescence, of our practitioner participants precisely because it adopted a format that was unlike that typical of social scientific study (that had precipitated some of the grievances discussed above).

Amongst the 'energy communities'
Once we had recruited and started to visit the various groups, it became evident that what we had naively taken to be a homogenous and united set of interests, albeit characterised by various local concerns, was to an extent 'shaped' by a competition amongst communities for fairly limited funding. For example, the LCCC was run by the Department of Energy & Climate Change (DECC), with twenty-two winners from an initial set of more than 500 applications. As a spokesperson for one of the eventually successful communities noted, this was an extremely arduous process entailing, amongst other things, a comprehensive rebuttal of an initially negative decision, and visits to contacts at DECC to lobby for a successful outcome. It was also clear that competition winners had successfully to organise the activities of their group in order to build effective and constructive networks that encompassed not only policy representatives, but also local entities such as councils, charities, and school groups, along with local and national businesses (e.g. infrastructure contractors). In this way, those with LCCC funding considered that their energy networks offered funders a framework that demonstrated capability in delivering a proposal.

In contrast, communities that were more ad-hoc and lacked organisational features, for example those that did not establish a programme of activity, nor formalise finances through the incorporation of a 'Ben Comm' (Industrial and Provident Society for the Benefit of the Community) or other instrument of legal constitution, were not well placed to make successful proposals. However, these informal groups

shared with the more coherent energy communities an underlying commitment to the Transition model (Hopkins, 2011). They therefore identified with a programme for delivering sustainability (of which renewable energy is seen to be a subset) that included the formation of a 'core group', 'partnership building', and 'community engagement', along with the planning and legal aspects mentioned above. Additionally, through identification with a common and what was seen to be urgent programme of sustainability, though relations amongst groups through their interactions with funding bodies including DECC were ostensibly shaped along competitive lines, the motivating context of the Transition network is in contrast represented as being supportive and mutual.

It is clear then that differentiation amongst the groups we researched is expressed through the extent to which they have made good the ambitions of Transition within their respective communities, which can on one hand be seen as the formalisation of shared commitments, while on the other hand becomes judged in relation to competitive frameworks by external agencies such as DECC.

One effect of success (as an index of competitive fitness) was the capacity for a community to be researched. Well-organised groups were able to contact, host, organise, and inform us more capably than ad-hoc groups. In this respect, our investment as researchers has principally concerned the LCCC winners – including time spent meeting, travelling to and from, speaking about, and designing in relation to. Well-organised groups have 'benefited' from exposure to design processes, including discussions about the themes of the workbooks, and descriptions of the technical platform and its features. Consequently, the successfully funded groups could become successfully researched, in that they were aligned with the research devices, having anticipated use and made schemes for the adoption of the final design. Finally, their experience of our research has allowed the successful groups to further expand their own plans, a point that is developed below in the discussion of communities enacting research.

Within the 'energy communities'
Despite variations in the 'effectiveness' of the groups we researched, we have argued that commitments such as Transition were often shared across the groups. However, those underlying principles were not necessarily shared within individual energy communities. Ostensibly, our energy communities were a set of diverse constituencies that included the populations of city areas (Bow and New Cross in London, the Meadows in Nottingham), village parishes (Ladock and Reepham), a city (Hastings), a geographic region (Sid Valley), and a university campus (Goldsmiths). We were, in reality, not dealing with 'whole communities' but rather with what social scientists would call 'samples', and these were highly structured by others, rather than by our own sampling strategy. Initial contacts were often key spokespeople within the communities, including those who

were doubtful about academic research, thinking (understandably) that funds should g

had led the formation of 'Ben Comms' and acted as authors of funding proposals. At other times our contacts were visible advocates of Transition who had arranged meetings within their communities and published those arrangements on blogs or through Twitter. These contacts then served as gatekeepers, and it was through them that we met others; and through these arrangements a particular and partial version of the energy community became reproduced.

The partiality of these arrangements was of course evident to the spokespeople, and their missing publics were considered in various ways. Representatives of ad-hoc groups at times agonised about the uptake of their projects by the wider constituencies in which they operated. More formalised groups had engagement strategies, and undertook events within schools and civic halls to enrol their publics. Groups also looked to other entities for support, for example legislative frameworks such as the Green Deal were anticipated as methods not only for generating income but also for recruiting members. Certainly, their publics were the direct beneficiaries of the energy-saving measures that had been resourced by the LCCC grants, where solar heating and photovoltaics were installed in homes, schools, and pubs. These infrastructures acted to bind and extend the core groups within their communities, and these measures were at times reviewed enthusiastically by their beneficiaries in terms of savings and features. However, at times the fiscal and utilitarian enthusiasms of the energy publics did not necessarily align with the underlying commitments of spokespeople, evidenced by the various ways in which the energy communities were composed.

Communities enacting research and enacting communities
From the outset of our engagement, community spokespeople were already enrolling publics through a range of activities supported through LCCC and other initiatives. Within this context, it was also apparent that the research project and the Energy Babble were construed as an additional means through which advocates were enacting communities. In this respect, how did the engagement of these groups with our project and researchers reinforce existing ties within the community, and what sort of representation of community became thereby enacted? How did the Energy Babble deployments provide a way of enrolling others, and how did Babble use differentiate or align communities?

Engagement by advocates with the project and with individual researchers were at times prospects for reinforcing existing ties within communities. So, while researchers identified community meetings as constructive occasions for fieldwork, requests for research visits were also taken as opportunities for convening social events for community members. A dinner party was convened to support one Babble deployment. This was an occasion of elaborate hospitality that supported the preliminary impact of our devices within the community and also supported the identity and affiliations of the group: at times explicitly, for example,

through after-dinner speeches. The research collaboration also provided the basis for enacting successful communities. Tours of LCCC-funded outcomes that were ostensibly organised for researchers were also occasions of celebration within the core group, and performances of accomplishment for the benefit of partners including local authority and county council representatives.

The plans and actions of individual adopters of the Energy Babble were diverse, though frequently the device was identified as a mechanism for enrolling others into an energy community. It was seen that the device could be taken to public events as a point of curiosity and as a catalyst for conversations, which would provide the basis for a membership drive. Babble was installed in schools where it could be used to broadcast data about savings derived from measures installed with LCCC support, and also for messages from groups which had been formed in order to report on cases of energy misuse. Here, the generation of audio content is an outcome of enrolment activities enabled by the device, and the playback of this material serves as a method of inscribing the community at the site where the Babble was installed, in this case the school library.

If practitioners deployed the Babble as a means of realising community of one sort or another, researchers also envisaged the Babble as affecting the communities. Specifically, we imagined that the device might be used to differentiate from, or align with other communities. For example, in workbooks – and in response to the competitive posture of groups in relation to scarce funding – scenarios were imagined where our design would be used as the basis for energy-saving competitions between communities. Imaginative tropes such as this, while often not borne out in use, demonstrate that the enactment of community and research was a complex, distributed exercise, undertaken by the researchers as much as the spokespeople for the groups.

Researchers enacting research and enacting communities
Our research sought to be driven by empirical encounters with the communities, both at the sites of their energy projects and at events where we undertook activities together. However, a great deal of our time was spent in our studio where we planned those encounters, interpreted resulting data, generated scenarios, proposed schemes, and made designs. Therefore, it is useful to consider our own hand in enacting research and enacting communities. In what various ways have we approached, and represented the ECDC project to diverse communities differently? How did we homogenise the various communities? Finally, how should we reflect on the value we hoped to have for the communities?

The presentation of the research project to the communities, and thus the enactment of those communities, varied in a number of ways. For instance, the articulation of the project in relation to communities developed over time, initially expressing community in relation to the DECC competition, then emphasising the identity of location

to direct efforts to reduce energy. So as the Babbles left their home in the studio and

and anticipating the features of the constituents, and later through specific engagements with spokespeople. Further, the settings in which there was a presentation to a community member might affect which features of the research were stressed: where slides shown to groups at their meeting in a hotel function room might have dwelt on a description of the research aims, a conversation during a journey between sites on a tour of carbon-saving measures might include reflections on funding schemes and project deadlines.

Here, we see how the informal and incremental features of design research served to perform contradictory, diverse, and partial versions of the communities. By contrast, the final execution and delivery of a range of finalised research devices and events arguably supported the homogenisation of community. In particular, the probe workshop, the probes, and the final design of the Babble can be understood in this way. These formalised formats act to resolve and anticipate their putative communities.

Concluding remarks

It is not easy to summarise these multiple enactments by multiple researchers and communities. What we can note by way of abstraction, is that what a community, or research, 'is', is subject to considerable variation as we move across different communities, different encounters, and engagements, and different phases in the research process.

Along the way, this variation is proliferated or reduced – depending on practical and rhetorical circumstances, communities, and research becomes more or less homogeneous, more or less differentiated.

In terms of the present ECDC project, this means that we need to tread carefully when we speak of 'communities', but also of 'research'. More specifically, to reiterate, we need to address in more detail how our research process was mobilised by our practitioner participants – how it fed into their local projects and processes. However, ironically, this is simply more of the same: each 're-newed' research engagement becomes yet another opportunity for those who are engaged to mobilise that 're-newed' research process to their own ends. Conversely, our 're-newed' engagement points to another prospect of variously proliferating and homogenising communities. This suggests that there is no 'end' to engagement – or rather, we need to think of it in radically dialogical terms, within a processual unfolding of the mutual constitution of communities and research. This point applies no less to the PEST literature – PEST initiatives are themselves appropriated and mobilised by publics; PEST itself examines how 'participation' and 'engagement' are enacted by publics. What the present discussion has hopefully done – at the very least – is identify and throw into relief this complex, mutualist iterativity of communities/publics and PEST/design research.

entered new worlds — homes, schools, pubs, and community spaces — they were almost inevitably

References

Arksey, H., *RSI and the Experts: The Construction of Medical Knowledge*, (London: UCL Press, 1998).

Bucchi, M., and F. Neresini, 'Science and Public Participation', in E. J. Hackett, O. Amsterdamska, M. Lynch, and J. Wajcman, eds, *The Handbook of Science and Technologies Studies* (Cambridge, MA: MIT Press, 2008), pp. 449-72.

Callon, M., P. Lascoumbes, and Y. Barthe, *Acting in an Uncertain World: An Essay on Technical Democracy* (Cambridge, MA: MIT Press, 2001).

Chilvers, J., 'Deliberating Competence: Theoretical and Practitioner Perspectives on Effective Participatory Appraisal Practice', *Science Technology & Human Values*, 33(2) (2008): 155-85.

Chilvers, J., and M. Kearnes, eds, *Remaking Participation: Science, Democracy and Emergent Publics* (London: Routledge, 2016).

Clark. T.,'"We're Over-Researched Here!" Exploring Accounts of Research Fatigue within Qualitative Research Engagements', *Sociology*, 42(5) (2008): 953-70.

Epstein, S., *Impure Science: AIDS Activism and the Politics of Science* (Berkeley: University of California Press, 1996).

Dunne, A.,. *Hertzian Tales: Electronic Products, Aesthetic Experience, and Critical Design* (Cambridge MA: MIT Press, 2005).

Dunne, A., and F. Raby, *Speculative Everything: Design, Fiction and Social Dreaming* (Cambridge, MA: MIT Press, 2013).

Felt, U., and M. Fochler, 'Machineries for Making Publics: Inscribing and Describing Publics in Public Engagement', *Minerva* 48/3 (2010): 219-38.

Gaver, W, A. Boucher, A. Law, S. Pennington, J. Bowers, J. Beaver, J. Humble, et al., 'Threshold Devices: Looking Out from the Home', Proceedings of the *26th Annual SIGCHI Conference on Human Factors in Computing Systems*, Florence, Italy (New York: ACM Press, 2008): 1429-38

Hagendijk, R., and A. Irwin, 'Public Deliberation and Governance: Engaging with Science and Technology in Contemporary Europe', *Minerva*, 44 (2) (2006): 167-84.

Hopkins, R., *The Transition Companion: Making Your Community More Resilient in Uncertain Times* (Totnes: Transition Books, 2011).

Irwin, A., and M. Michael, *Science, Social Theory and Public Knowledge* (Maidenhead, Berks.: Open University Press/McGraw-Hill, 2003).

Kerridge, T., 'Designing Debate: The Entanglement of Speculative Design and Upstream Engagement' (PhD Thesis, Goldsmiths, University of London, 2015).

Lezaun, J., and L. Soneryd, 'Consulting Citizens: Technologies of Elicitation and the Mobility of Publics', *Public Understanding of Science*, 16 (2007): 279-97.

Marres, N., *Material Participation: Technology, the Environment and Everyday Publics* (Basingstoke: Palgrave, 2012).

Michael, M., 'Publics Performing Publics: Of PiGs, PiPs and Politics', *Public Understanding of Science*, 18 (2009): 617-31.

— '"What are we busy doing?": Engaging the Idiot', *Science, Technology and Human Values*, 37(5) (2012): 528-54.

— 'Notes Toward a Speculative Methodology of Everyday Life', *Qualitative Research* (forthcoming).

Michael, M., B. Costello, I. Kerridge, and J. Mooney-Somers, 'Manifesto on Art, Design and Social Science – Method as Speculative Event', *Leonardo*, 48(2) (2015): 190-91.

Michael, M., and W. Gaver, 'Home Beyond Home: Dwelling with Threshold Devices', *Space and Culture*, 12 (2009): 359-70.

Nowotny, H., P. Scott, and M. Gibbons, *Re-Thinking Science: Knowledge and the Public in an Age of Uncertainty* (Cambridge: Polity Press, 2001).

Papanek, V., *Design for the Real World*, 2nd edn (London: Thames and Hudson, 1984).

Storni, C., T. Binder, P. Linde, and D. Stuedahl, 'Designing Things Together: Intersections of Co-Design and Actor-Network Theory' *Co-Design*, 11, (3-4) (2015): 149-51.

Telier, A., *Design Things* (Cambridge, MA: MIT Press, 2011).

Wynne, B.E., 'The Public Understanding of Science', in S. Jasanoff, G.E. Markle, J.C. Peterson, and T. Pinch, eds, *Handbook of Science and Technology Studies* (Thousand Oaks, CA: Sage, 1995), pp. 361-88.

camed with attributions of utility, and portrayed as new sources of information and

Reepham Life, November 2013

environment

Energy Babble will help Reepham reduce energy

REEPHAM is one of the UK's first towns to trial a prototype of new technology designed to engage communities in reducing energy consumption. The project addresses how to achieve an 80% reduction of the country's carbon emissions by 2050.

The "Energy Babble" has already been installed at both schools in Reepham, and units are planned for other venues in the town.

Designed by Professor Bill Gaver and his project team at Goldsmiths, University of London, the Energy Babble collects information relating to energy issues from an extensive network. A server process aggregates and transforms input sources into audio files that are broadcast by each device.

The idea is to open and promote constructive debate and involvement on energy reduction issues. Anyone with access to a Babble can send information to it, either verbally using the microphone or remotely via texting or the internet. The information is then logged, checked, stored and broadcast within a few minutes.

Goldsmiths was awarded a £795,000 grant to fund the Energy Babble over a three and a half year period.

Prof. Gaver said: "The most fundamental achievements will be the Babble on the one hand, and people's engagements with it and reactions to it on the other.

"The Babble itself explores technical

Matthew Plummer-Fernandez of Goldsmiths, University of London, with one of four Energy Babbles being installed in Reepham

possibilities, but also summarises, in a sense, the situation of people who are trying to make progress on environmental action at a community level.

"The reactions don't just tell us whether people like the Babble, but also – we hope – will reveal a lot of their knowledge, concerns and beliefs about environmental issues, as the many different ways people minimise their energy consumption."

He continued: "The Babble is not a blank slate like Twitter and other social media, but is far more centred on environmental issues. Also, because the Bab-

ble is a physical device it has a presence, and a social one, that on-screen systems like Twitter, etc., may not have.

"Using audio output means it is more pervasive; you don't have to look at it to engage with it. It is designed to highlight the ways people talk about environmental issues, and that is more or less all it does, all the time."

As a winner of the Low Carbon Communities Challenge, the Reepham Green Team were introduced to Goldsmiths two years ago to help develop the Energy Babble.

In total, 35 prototypes have been made available to trial nationwide, of which four units have come to Reepham, two of which are already installed in the schools.

Judy Holland

■ The Green Team would like *Reepham Life* readers' help in deciding where the other units should be installed. Please send your suggestions to: info@reephamlife.co.uk, or in writing to Reepham Community Press, Homerton House, 74 Cawston Road, Reepham, Norfolk NR10 4LT, or left at Very Nice Things in the Market Place.

www.reephamchallenge.org

www.reephamlife.co.uk

The arrival of Energy Babble featured in Reepham Life

communication among communities. To be sure, they did provide information and broadcas

Using

The idiosyncratic appearance and odd transmissions of the Babbles were often met with some puzzlement. During the deployment meetings, people interpreted the devices according to their own interests and desires, for instance seeing their potential use in publicity events or as a means to publicise their successes. In return, we found ourselves glossing the devices as providing a source of information to the communities, combined with a means of communication amongst them. Such a characterisation helped us reassure the communities about the Babbles, but when we returned months later it seemed that this might have confused peoples' understanding of the devices.

CLIP 128

Duration 3.53 Minutes

JINGLE CEMBALO CHORD 1

Announcer: Allison (american english, female voice)
When I turn on the elections in the government

JINGLE KEYBOARD 1

Announcer: Ava (american english, female voice)
Recent message: Looking forward to febubabble

SOUND: RUMBLING NOISE

JINGLE KEYBOARD 2

Announcer: Kate (british english, female voice)
...saving lives sustainble transport

JINGLE PETER AND THE WOLF

ANNOUNCER: Oliver (british english, male voice)
Do you have a problem with coal?

JINGLE CEMBALO CHORD 1

Announcer: Allison (american english, female voice)
... Does anyone know power rating for the energy babble?

JINGLE GLOCKENSPIEL

Announcer: Jill (american english, female voice)
Omg something about imperialsm and energy crises

Transcript of a 3-minute broadcast

peoples' words, but not in the well-behaved and helpful way people might have expected.

```
                              -----------
                              ----------
                              0
                              837000
008000          pi08
                (?) High School has
, !, ,          their own winter (?).
                -----------------------
-----------     2013-12-10
----------      11:16:57.433000
7               vox
302000          pi08
                Reepham High School
our             has it's own wind
log by          turbine.
s name. Are     -----------------------
know.           2013-12-10
-----------     11:20:41.177000
7               sms
741000          Does your Wind
                turbine have a name?
we interact     -----------------------
                2013-12-10
-----------     13:22:46.180000
---------       vox
)               pi25
191000          Happy Christmas.
                -----------------------
ght first       2013-12-10
ents.           13:23:08.556000
-----------     vox
----------      pi25
)               Alex says enjoy your
522000          day.
                -----------------------
                2013-12-10
                17:27:16.794000
-----------     vox
---------       pi25
)               This is ECDC 31, we
050000          are signing in.
                -----------------------
name is         2013-12-10
ase respect     17:28:54.323000
                sms
-----------     Hello ECDC 31. We
----------      hear you loud and
)               clear on our babble
499000          -----------------------
                2013-12-10
nment.          17:29:07.600000
----------
```

```
I like cookies. Yes
it is getting cold
---------------------
---------------------
2013-12-10
17:34:55.193000
vox
pi25
We live in less rooms
now, because some
have no insulation at
all.
---------------------
---------------------
2013-12-10
17:36:15.657000
vox
pi25
The world has many
floods but less talk
of the climate.
---------------------
---------------------
2013-12-10
17:37:02.043000
vox
pi25
More floods, less talk
of the climate.
---------------------
---------------------
2013-12-10
18:04:41.139000
vox
pi25
At home, that is.
---------------------
---------------------
2013-12-10
18:05:01.434000
vox
pi25
In the weekend
Twitter use goes
down, heating use
goes up.
---------------------
---------------------
2013-12-10
18:11:48.366000
vox
pi25
How do we interact
with this.
---------------------
---------------------
2013-12-10
18:34:02.760000
vox
pi25
(?) doesn't speak
Dutch, its (flur's?)
birthday.
---------------------
```

```
that has on energy
demand.
---------------------
---------------------
2013-12-11
14:16:55.396000
sms
Solar tracking
technology sounds
like a really good
way to get the most
out of a solar panel
---------------------
---------------------
2013-12-11
15:21:35.358000
sms
Heating is always on
in the flat beneath
us. They are council
tenants, i wonder
if they have their
heating bill paid
for?
---------------------
---------------------
2013-12-11
22:05:43.110000
sms
Hi Pete
---------------------
---------------------
2013-12-12
08:31:55.815000
vox
pi14
The revision in
government strike
prices. Shows a lack
of commitment.
---------------------
---------------------
2013-12-12
08:59:03.547000
vox
pi25
(Paddles?) are
expensive, if you
rent your house.
---------------------
---------------------
2013-12-12
09:21:14.001000
vox
pi25
Hello Babble, this
morning we hear you
loud and clear.
---------------------
---------------------
2013-12-12
09:24:27.065000
vox
pi25
```

**Log of contributions from users
via microphone and SMS**

d and
    getting

Instead, snippets of news jostled against bits of nonsense, people's contributions via the

-13
6.785000

rning, (say
n), good
. Good morning
dy.

-14
9.667000

h power has
ed a five
our billion
lan to build
lds largest
e wind farm
 coast of

-14
2.445000

groups have
ned that
es available
 wind projects
cottish power
s they plan
d a off shore
rm near Tyri
 announced 2
go that it was
ing plans to
ct a 4 billion
ind farm of

-16
9.804000

is not
med to
se jobs but
 pick one up
search for
t news.

-16
5.226000

 the stunt
by the
g protestors
e wind turbine

----

part of the mainland
grid as the output is
not regular.

2013-12-18
08:40:03.783000
sms
RWE yesterday got
consent for its 36
mega watt onshore
wind project in
brechfa forest,
Carmarthenshire. This
is adjacent to its
larger 84 mega watt
scheme in brechfa
forest west. A great
result for onshore
wind in Wales.

2013-12-18
08:43:46.120000
vox
pi14
{unclear} the
proposed nuclear
power station at (?).

2013-12-18
11:58:20.757000
sms
Sidenergy has now
been launched

2013-12-19
12:51:21.604000
sms
Merry christmas
babble

2014-01-01
12:11:53.353000
vox
pi25
Happy New Year.

2014-01-01
12:12:13.827000
vox
pi25
They can sepak into
the mic as well.

2014-01-01
12:13:34.247000
vox
pi25
Thi...

23.28.48.423000
vox
pi21
Why do British Gas
charge me 70 Pounds
for electricity when
I am in credit for 70
Pounds in my gas?

2014-01-05
15:27:57.787000
vox
pi21
I've just submitted
my (fee?) in tariff.

2014-01-05
23:22:23.792000
vox
pi21
I do find Winter
phones quiet
beautiful.

2014-01-06
23:55:23.561000
vox
pi21
I'm looking forward
to my energy babble
and wifi.

2014-01-09
11:43:37.669000
vox
pi21
The generated point
of 8 of the kilowatt
hour yesterday.

2014-01-09
13:11:01.890000
vox
pi25
And my colleagues
in the Netherlands
listen to babble
online.

2014-01-09
13:17:04.861000
vox
pi25
 That was a
question. Does
babble answer
questions.
----

carbon emissions in
the US raised two
percent. Does one
outweighs the other?

2014-01-15
11:54:46.590000
sms
Energy reduction and
January don't go wel
together, lots of
guilt

2014-01-15
15:45:21.950000
sms
The geezers presente
a fantastic project
at the white buildin
in hackney wick
yesterday, well done
geezers from bow

2014-01-15
17:35:50.495000
sms
Does photo voltaic d
much in winter?

2014-01-16
07:04:40.444000
sms
Huge biomass boiler
in hackney wick on
the Olympic site.
Right next to the
canals and barges
that use wood burner
for hot water. Into
hot water.

2014-01-16
08:33:33.197000
vox
pi07
Global greenhouse ga
emissions are set
to rise by nearly a
third in the next 2
decades a new report
by DP has found.

---

l-Daniel:
ardian says: World leaders
st now respond to an
equivocal message from the
w L. E. Ds. Have you taken
em off.

'S-Serena:
ardian Environment says.
ark cull: Greg Hunt exempts
. from laws protecting species
 risk

14-01-21 06:03:32.008245

.Q-Lee:
at is the problem with
sics? so, what do you think
out system?

R-Serena:
cent Message. Good morning
eda, how are you doing
day?. Aha. morning. How do
u feel about that?

.Q-Lee:
w does electricity relate
 gas? do you have a problem
th heat? does anyone have any
oughts on cell?

---

l-Daniel:
turn on the Olympic site.
ght next to the ? college
morrow for the cold winds.

14-01-21 06:04:57.007022

---

l-Daniel:
llo, hello, hello, hello,
llo, hello, hello. the
oposed nuclear power station
 Vanalac? . We may have
st the recognition or just
e ability to go through the
mputer. I am in credit for 70
unds for electricity when I
 testing ECDC three which is
ving fuel.

.Serena:
nt message: Frida is a good
 for a babble, my cat is
ed frida..

2014-01-21 06:06:50.007937

---

bM-Daniel:
Testing bagel for grinning gold
smiths.

---

bTS-Serena:
BBC Radio 4 Today says. This
mornings provisional running
order is available here:

---

bTH-Serena:
Solar PV was a heatwave saviour
Business Spectator. Alan
Kohler is one of Australias
most experienced commentators
and journalists. Alan is the
founder of Eureka Report,
Australias most successful
investment newsletter, and
Business Spectator, a 24-
hour free business news and
commentary website. He also
hosts Inside Business, a half-
hour Sunday programme on the
ABC, is the finance presenter
on the ABC News - and producer
of the nightly graph or two.
Speculation mounts that Shells
Woodside stake will be next to
go, while Air New Zealands boss
joins the Virgin board.

---

bEG-Serena:
the U K energy demand has gone
down by 549 mega watts.

2014-01-21 06:08:16.007249

---

bTA-Serena:
Essential Report . Q. If a
Federal Election was held
today to which party will
you probably give your first
preference vote? If not sure,
which party are you currently
leaning toward?

---

bRR-Serena:
We recently heard a listener
say. Frida is a good name for
a babble, my cat is called
frida.. Yes, well said, we
should discuss the name

---

bEG-Serena:
the U K energy demand has gone
up by 415 mega watts.

2014-01-21 06:11:59.008668

Today says. :
 r4today Tuesday
eneral Stanley
alks Afghan
ith

---

bTE-Moira:
DJ McKenna from says. 50 Coal
burnin power plants in the good
ol USA, what say you HYPOCRITE?

---

bTA-Serena:
Tenesol Solar Panels Tenesol
Solar Panel 185W in South
Africa.. The use of Solar Panel
Energy is not limited to only
providing an avenue of safe
energy production, there are
actual fiscal rewards tied up in
producing your own sustainable
energy supply for your home.

---

bRR-Serena:
describing that as energybabble
is funny

2014-01-21 06:12:26.006608

---

bTE-Moira:
Joe from says. EU and UK
reason our energy cost twice
that of the US? environmental
cost, green taxes! price rise
on all fuel bills, transport of
goods food!

---

bRR-Serena:
We recently heard a listener
say. I have an energybabble and
very lovely she is too. Yes
energybabble!Recent Message. I
have an energybabble and very
lovely she is too. So, very
energybabble

---

bM-Daniel:
Hello this is the most out of a
British smell, sorry to bring
down British Energy bills. Okay
Vox, Im just checking ECDC 2
from ? college tomorrow for the
cold winds.

---

bTA-Serena:
Indias coal imports rise 20
pct to help fuel new power
plants Reuters . NEW DELHI
Jan 7 Reuters - Indias coal
imports rose 20 percent to
105.8 million tonnes in April-
October from a year earlier
as power producers turned to
Indonesia to help feed new
plants, according to data from
mjunction services, an online
market operator. Regulatory
and bureaucratic delays in
adding new mines and expanding
existing ones have made India
the No. 3 importer of coal,
even though it sits on what
BP ranks as the worlds fifth-
largest reserves. Imports
leaped 34 percent to 137.56
million tonnes in 2012/13.

2014-01-21 06:14:40.013052

---

bTE-Moira:
Maahes. from Karnak. says.
People are so rarely worth the
energy they demand.

BBC Radio 4 Today says. (
up: Funeral costs 0650, w
next for Afghanistan when
troops leave? 0710, and
of the apology 0820

---

bRR-Serena:
Recent Message. I have an
energybabble and very low
she is too. Yes I agree,
is energybabble

---

bTH-Serena:
Indias coal imports rise
pct to help fuel new power
plants Reuters . NEW DELH
Jan 7 Reuters - Indias co
imports rose 20 percent
105.8 million tonnes in A
October from a year earli
as power producers turned
Indonesia to help feed ne
plants, according to data
mjunction services, an on
market operator. Regulato
and bureaucratic delays i
adding new mines and expa
existing ones have made I
the No. 3 importer of coa
even though it sits on wh
BP ranks as the worlds fi
largest reserves. Imports
leaped 34 percent to 137.
million tonnes in 2012/13

2014-01-21 06:16:36.02000

---

bTH-Serena:
Solar PV clipped peak dem
by 4.6pct during heatwave
Renew Economy. Victoria a
South Australia have just
through a week of very hi
temperatures and very hig
maximum electricity deman
There has been some debat
to what contribution if a
solar PV has made. Our an
shows that solar PV has m
a significant contribution
being responsible for red
peak demand by 4.6 per ce
According to electricity
demand data published by
Australian Energy Market
Operator AEMO the peak
electricity demand occurr
Thursday 16th January dur
the half hour commencing
pm in Victoria 10,240 MW
during the half hour comm
6.30 pm in South Australi
3,246 MW. South Australia
Victoria are interconnect
so to properly assess the
contribution of solar PV
have considered electrici
demand and PV contributio
across both states.

---

bEG-Serena:
the U K energy demand has
up by 981 mega watts.

---

bTE-Moira:
Tanya Ha from Melbourne s
This Ron Tandberg cartoon
of sums up Oz blindness t
renewable energy opportun

**Extract of broadcast showing the voices
of the Energy Babble. Each voice is
constructed by a specific algorithm.**

Design and Science & Technology Studies

Mike Michael

Introduction

In recent years, design has been particularly interested in one branch of social science, namely science and technology studies (STS). In some ways, this is not unexpected – after all, design, broadly put, has entrée into a variety of technological and scientific endeavours (e.g. synthetic biology, industrial and product design, information technology, the built environment, vehicle design, etc.) and is thus ripe for STS analysis (e.g. Strengers 2013; Wilkie 2013). Design is thus another technical discipline that can be subjected to forms of STS analysis – analyses which ask, for instance, what assumptions go into the design of particular artefacts (sociological questions can also be posed in relation to the ways in which design is shaped by systems of production – e.g. Molotch 2003). Design also speaks to STS's preoccupation with everyday technology and the ways in which this does, or does not 'work'. Here, the longstanding STS interest in what counts as 'working' (ranging from successfully accomplishing local tasks to accidentally contributing to global climate change) maps onto design's own interest in how to make stuff 'work'. Thus, as writers such as Bijker (e.g. 1995) and Verbeek (e.g. 2005) have noted, design can help us understand how technologies can be built to incorporate new aesthetic, ethical, and political values.

In the first instance, design is a topic: it is simply one object amongst others that can be subjected to social scientific analysis. Here, the take-up of design as an empirical 'object' can be seen to correspond with recent 'turns' within STS, notably the turn to materiality (e.g. Marres and Lezaun 2011) and the turn to ontology (e.g. Woolgar and Lezaun 2013). In the second case, design is a resource – it can inform how social science goes about its business of analysing social processes. On this score, designers who provide alternative visions – of the world, of users, of how users serve in design procedures and practice – can serve as a means towards developing social critique.

If, however, we widen our scope and consider the relationship between design and social science more broadly, then what we see are much more intricate and intimate interdependencies that have played out in and over multiple genealogies. Such interrelations can be seen to inform, or at least provide a backdrop to, a fragmentary pattern of engagements between design and STS. This, for example, can be detected in the works of, for want of a better handle, 'academic designers'. Social science and STS are unlikely to have been of direct and discrete interest to designers. At best, they will have been 'addressed' by designers only by virtue of being involved in such generic elements of academia as pedagogy (e.g. the classroom, virtual teaching) and publishing (text layout, online article processing) that have ostensibly been subject to considerable influence by various design disciplines – architectural, graphic, product, interaction, etc. While this all treats social science as a topic of sorts (a weakened form of interdisciplinarity), there are other genealogies where social science has resourced design thinking, or perhaps more accurately, where the boundaries between design and social science are not so easily demarcated. Here, for instance, we can take note of the work of Otto and Marie Neurath at the turn of the twentieth century (Cartwright et al. 2008). As a foundational blend of statistical and design reasoning, it observes how Scandinavian participatory design actively sought to achieve democracy in the workplace by way of Marx and Wittgenstein and echoes of action research, and points to the influence of Marxist and proto-environmental social scientific concerns in the classic work and critical design work of Papanek (1984). Moreover, it is not unreasonable to detect echoes of 'action research' in participatory design, the take-up of ethnomethodological approaches in human-computer interaction design (e.g. Suchman 1987), and, more recently, the trend for ethnographic data to resource designing, and even in so-called 'speculative design' (e.g. Gaver et al. 2008). And it is not difficult to point to the impact of various social science writers (e.g. Latour 2005; the Frankfurt School) in variants of critical design (Dunne and Raby 2001, 2013; DiSalvo 2012; Wilkie and Ward 2008). Here, then, social science becomes a resource for design, impacting on the options – both conceptual and practical – that can become available to it.

However, against the idea of design as STS's topic or resource and vice versa, this essay counterposes the picture of design and STS as 'collaborators'. That is to say, we explore the interdisciplinary ways in which STS and design can work together, such as those mentioned above, such that together fashion new 'objects of understanding' (see Barry et al. 2008). So, in what follows, we discuss how design and STS can 'co-engage with an empirical field'. However, design and STS are each far too large as fields: we could not possibly do justice to them in so short an essay. Instead, we focus our efforts on those 'traditions' within design and STS which have entered into the project presented in this volume, namely, speculative design, and recent developments in 'actor-network theory' (or post-ANT). Nevertheless, a number of issues are raised, for what counts as 'engagement' and 'an empirical field' is not always shared.

In this essay, we look at several points of contact – ideally sites of mutual immersion – at which post-ANT and speculative design meet if they are to collaborate. These points are, in large part, pragmatically derived from the exigencies imposed by an institutionalised requirement to work together collaboratively. So, we can suggest points where negotiations and calibrations need to take place. Minimally, these might include: the crafting of 'the research question'; the choice of empirical sites; the empirical method/the process of field engagement; the analysis of data/the treatment of collected materials; and the presentation of data/materials and their analysis/ treatment. However, this makes things sound a little too straightforward. This is because behind these points lie much bigger issues concerning, for instance: methodological questions about the means by which 'the empirical' is to be engaged (which methods do we use and is the social world described, intervened in, or enacted through such methods?); epistemologically, there are differences in how empirical material is treated (do we seek 'patterns in the data'; or generate idiosyncratic insights that draw on an unsystematic variety of sources?); and there are divergences in the political and ethical expectations that attach to the researcher (put crudely, is our role one of 'critique' of a definable state of affairs, or a provocation of possibilities within a world that is unfolding?). Having made these points, we should not forget that collaboration also arises through more personal and cultural dynamics – these are touched upon in the 'A prehistory' anecdote.

In light of this, in what follows below we will begin with an overview of some recent discussions of design research, paying particular attention to the character of speculative design. We will then consider some of the recent discussion in social science, especially those branches that are STS-inflected (post-ANT) and concern a view of method as performative, and a perspective on the world as heterogeneous, unfolding, processual, and relational. We will then bring these together to discuss some delicate points of intersection. We end by situating the present project in relation to these.

Design toward the speculative

'Speculative design' can be placed within a wider historical trajectory that arguably takes in such traditions as 'participatory design' (in which the process of designing objects or systems is directly shaped by the invited input of potential and future users so that the designs better serve users both practically and politically – see Ehn 2001) and 'critical design' (in which designers project sometimes believable, but often provocative, technological futures in order to design artefacts and/or systems which can serve to critique those futures by posing questions about what those futures suppress or assume – see Dunne 2005; Dunne and Raby 2001). In addition, it can be suggested that critical and speculative design can be related to architectural genealogies, notably experimental and radical architectural design exemplified by the work of Superstudio (see Lang and Menking 2003), Cedric Price, Constant Nieuwenhuys and Rem Koolhaus, amongst others. It goes without saying that the borders between these traditions are highly porous. For instance, a recent example of this porosity is the work of DiSalvo (2012) which develops an approach he calls 'adversarial design', which draws on and combines elements of both participatory and critical design. Latterly, practitioners of participatory design have proved adept at both incorporating the insights and problematique of elements of STS while also developing their own 'speculative sensibility' (e.g. Binder et al. 2011, 2015). And Dunne and Raby (2013) have, again recently, explicitly highlighted how 'the speculative' informs their own design practice as well as that of others. While the 'speculative design' that characterises the work of the Interaction Research Studio and its members (e.g. Boehner, Gaver, and Boucher 2012; Michael and Gaver 2009; Gaver et al. 2008; Wilkie, Michael, and Fernandez 2015) bears some resemblance to, and shares some common references with, participatory and critical design, it also displays some unique elements. Indeed, members of the Interaction Research Studio voice unease when it comes to characterising the work as 'speculative design'.

Having said this, a certain amount of care is required when it comes to demarcating what counts as speculative design, writing histories of design practice, or for that matter, delineating the interrelations between design and STS. Indeed, even what we describe as the 'Interaction Research Studio' comprises a shifting and evolving set of expertise and interests out of which the ECDC project emerged. Thus, the project team that comprised the ECDC project brought together interests in, and longstanding commitments to human-computer interaction design, product and industrial design, visual and graphic design, and STS.

purposely undermined its own utility, reflecting the coming together of a design practice

The 'speculative design' that has been developed by the Interaction Research Studio can be differentiated from other forms of design, including some to which the term 'speculative' is also attached, along a number of parameters. For instance, in relation to the role of the user in the design process, user input is both more and less 'mediated' than it is in more traditional forms of participatory design practice. While in the latter users engage in the evolution of prototypes by working directly with them, in speculative design potential users supply 'material' that feeds into the design process. This material can be ethnographic. This is relatively 'unmediated' insofar as the designers collect material through ethnographic visits to the domestic, work, or community settings of the potential users. This might involve informal conversations, non-participant observation, or the production of photographic records. At the same time, speculative designers also gather more 'mediated' material through the highly 'artificial' means of cultural probes (see below). As noted elsewhere in this volume, cultural probes are concerned with playfully encouraging participants to reimagine their relation to their social and practical settings (in order to tap into the ways in which such settings might unfold). Such probes vary in what they ask of their users (e.g. soliciting views on the aesthetics of a dwelling's energy use, or providing the opportunity for idle doodling while talking on the phone, or enabling participants to take photographs of, say, a home's spiritual centre), but they all aim to provide a sense of the ambiguous, immanent, tangential dimensions of aspects of participants' lives.

Focusing on users also draws out further distinctions with speculative design exemplified in the work of Dunne and Raby. As mentioned, who 'users' or participants (to use the nomenclature of the Studio) are, and what they are capable of, emerges through the process of design, which often points to empirically near or 'proximal' futures. Here, however, the 'users' tend not be configured but rather remain as indistinct actors capable of ambiguity, irony, etc.

Having drawn these comparisons, as we shall see below, mediation and artificiality are highly problematic categories – however, for present purposes they serve to throw into relief the differences between participatory and speculative design. Another difference lies in how speculative and participatory design make use of their participants' contributions. Participatory design makes direct use of such inputs – participants are thus quasi-collaborators in the design specification of artefacts and systems. For speculative design, the materials collected from participants again play a much more 'mediated' role in the design process. Specifically, these materials are set within a whole array of other sources, data, images, histories etc. They are combined – in an opportunistic and individuated way – with, for instance, fragments from design and art history, magazine articles, technical papers, official policy documents, collections of relevant technologies. These heterogeneous materials are also collected and 'edited'

in the form of workbooks which serve as common foci for developing design ideas (Gaver 2011). The intermediary result of this (part of) speculative design practice is a 'brief' in which the key parameters of the design are delimited.

There is a reverse comparison to be drawn between speculative and critical design. If speculative design tends towards relatively mediated engagements with participants (when contrasted with participatory design), there is nevertheless engagement per se (unlike much critical design). To be sure, both speculative and critical design enact a certain sort of 'designerly mystique' insofar as it is the designers themselves who ultimately develop the design (even if in the former case, participants have contributed). However, there is a key difference in the designerly 'use' of designs. For speculative design, the artefact and/or system must 'work' because they are deployed – sent out into the world to prompt yet more ambiguity and playfulness, and to further probe the not-as-yet that attaches to the specific social and practical worlds of users. This links to another round of material collection through ethnographic visits. For critical design, the 'concept' of – the idea behind – the critical design takes precedence over its actual functioning: insofar as it inspires critique of the related technological future, there is no special need for deployment, only access through exhibitions or published texts.

In this section, we have attempted to differentiate speculative design from two of its closest relatives – participatory and critical design. We have done this mainly by considering the different design practices that characterise each (e.g. the involvement of the user). However, there are also divergences around the 'point' of design. Design has traditionally been concerned with the future. However, how such a future is related to the present varies considerably across design genres. For straightforward product design, say, there is a proximal and linear relation between the present and the future: contemporary problems deserve near-future solutions and product design will provide those. Of course, things are more complicated: after all, this apparently linear movement entails projections of future and present users, contexts of use, shaping of futures and future users in the present in order to realise futures in which solutions are indeed solutions (as the sociology of expectations – a sub-branch of STS – makes clear – e.g. Brown and Michael 2003). For participatory design, the future and the present can be said to be intimately intertwined through the process of doing design. The participation of users not only shapes the product, but also models a more democratic design process, a model that ideally impacts on the politics by which problems are identified, solutions negotiated, and implementation accomplished (not least in the workplace). For critical design, as we have seen, credible technological futures are identified, and then critiqued by virtue of a design that in some way or other demonstrates the 'poverty' of that future. If, broadly speaking, standard design designates a positive functional future, participatory design aspires towards a positive sociopolitical future, and critical design articulates negative sociopolitical futures, then speculative design orients towards an 'open' future. Put another way, the objects and systems developed through speculative design (and that includes cultural probes) serve as means to evoke a 'proximal' future – the complex, ambiguous, playful unfolding of the present in ways that are not always expected and which, potentially at least, open up the possibility of reformulating what is at stake in the situation into which the speculative design is introduced. We shall have much more to say about this in a later section when we discuss how speculative design and STS might interdigitate. Before that, however, we need to consider work in STS (and in social science) which resonates most closely with speculative design.

STS, ANT, post-ANT

Science and technology studies covers a multitude of approaches and issues, though the particular tradition we draw upon here is that of the sociology of scientific knowledge (SSK). SSK has been particularly concerned to study how scientific knowledge is shaped by social practices, not least how putatively 'objective' knowledge about natural phenomena emerges through rhetorical and representational processes. At base, these processes involve scientists' efforts to accredit themselves and their supporters, while discrediting their opponents (i.e. those who hold to different accounts of the natural phenomena under disputation). Into this broadly social constructionist approach, actor-network theory (ANT) introduced a sense of materiality – the production of knowledge was shaped not only by the success with which scientists could recruit supporters to their epistemic cause, but also by the role of nonhumans. Thus, nonhumans such as microbes, electrons, wind patterns, and so on could all be treated as 'actors' (or 'actants') that might resist or assist in the establishment of accredited scientific knowledge. If scientists could 'translate' nonhumans and humans so that their roles aligned and their actions 'converged', they could build actor-networks in which all these human and nonhuman actors worked seamlessly together as one. This 'agential' role of nonhumans was extended to the production of social relations more generally. Thus, mundane technologies such as key weights and door closers, by affording or denying particular actions, were understood to be central to how people might associate with one another. Those who had a hand in designing such technologies could thereby impact on the sorts of associations possible amongst people, as well as between people and technologies (a small selection of classic references might include: Latour 1987, 2005; Callon 1986a, b; Law 1987).

To be sure, this is a very condensed overview of ANT (for an extended summary of this field, see Michael 2016), but hopefully it provides sufficient background to move on to more relevant post-ANT developments. ANT has been critiqued on several grounds, though in the present case we will mention just two of the most important for present purposes. Firstly, ANT has been too 'managerialist' or 'militaristic' – the picture of a central actor, such as a scientist or group of technologists, who 'enrols' or 'mobilises' others to form a network can do justice neither to the ways in which some actors are altogether marginalised, nor to the ways in which that central actor itself emerges out of complex sets of relations (that are not necessarily agonistic). Secondly, ANT is too 'empiricist': Latour's (1987) methodological injunction to 'follow the actor' (e.g. scientist) cannot address the ways in which social research is instrumental in 'making' the object that it is studying.

Amongst the various arguments that fall under the (wide) rubric of post-ANT (e.g. Michael 2016), we can point to the following. Instead of the metaphor of network with its (ontological) evocation of agonism and linearity, various writers have turned to metaphors of assemblage, fluids, rhizome, and topology in which connections between actors are multiple, non-linear, and emergent. And in contrast to the empiricism of early ANT, writers are increasingly interested in the performativity of method – how we as researchers play a part in 'enacting' the phenomena which we aim to study (e.g. Law 2004; Mol 2002). We see both of these dimensions of post-ANT brought together in John Law (2004)'s reframing of social scientific method as a 'method assemblage'. On the one hand, Law regards 'reality' in assemblage terms – a nexus of multiple, shifting, and emergent relations, a 'world of becoming' as Connolly (2010) frames it. As such, any particular entity arises out of a complex set of relations. At stake for the social scientific researcher is how to empirically grasp this shifting, emerging, processual array of multiple relations. The idea behind a 'method assemblage' recognises that when a researcher enters the research field (itself, of course, composed of flux, emergence, and multiplicity) she too takes part in the processuality of that reality. For Law to do research is to engage in active, uncertain, faltering ways with that reality while nevertheless simultaneously performing, enacting, or making that reality. This is because any such engagement is necessarily selective – only some elements of an assemblage are registered. As Law famously puts it, a method assemblage entails the 'crafting of a bundle of ramifying relations that generates presence, manifest absence, and Otherness' (Law 2004: 45). However, by taking this into account, the researcher can be more open to possibilities entailed in her 'use' of a method assemblage to the extent that she can 'imagine more flexible boundaries, and different forms of presence and absence' (ibid.: 85).

We can approach this way of thinking about method through the work of a post-ANT fellow traveller, namely Isabelle Stengers (e.g. 2005a, b). Drawing on the work of

Whitehead and Deleuze (who have both had some influence on ANT and its adaptations), Stengers focuses on the notion of event. Accordingly, events are, drawing on Whitehead (1929), 'actual occasions' in which a multitude of different sorts of entities (or prehensions) – entities that are, for instance, variously social or material, human or nonhuman, conscious or unconscious – collect themselves and combine (or concresce). In the version of the event we advocate here, we see these various elements or prehensions as mutually affecting one another, co-becoming as they concresce (see Fraser 2010).

Research events are, of course, a subset of events per se. As such, the process of researchers engaging empirically with the world not only affects that world, but conversely that world affects the researchers. In other words, both the 'object' of social research and the researchers themselves co-become – emerge together within the research event. One implication is that, given that the world and the researchers mutually change, it becomes problematic to continue to understand the event of their co-becoming as a 'research event'. As Michael (2012a, b) has noted, the research event becomes along with the co-becoming of its component elements: it becomes something other, for instance, an exercise in irony on the part of research participants, or an occasion for tidying up the 'mess', as Law puts it, of the research event in order to sustain the political and institutional commitments of the researchers. More positively, we can also see the uncertain unfolding of the research event as an opportunity. Rather than trying to 'solve' the constitutive indeterminacy of the research event by, for instance, ignoring its less alluring or comprehensible aspects (especially when research participants 'misbehave' in one way or another), we can take these into account and see the research event as posing more intriguing research problems (e.g. Wilkie et al. 2015). This means, as Fraser (2010) puts it, asking better questions or engaging in 'inventive problem-making': this entails a reformulation of the issues at stake that underpin the research event. But further, because in the process of the research event the researcher has also changed, we might ask who or what is doing the research – and is it any longer research? This all reflects the open, unfolding, emergent, relational character of the event. Given this chronic uncertainty of the event, we are left with much more leeway in how we grasp – that is, enact or 'make' – the research event (which may no longer have much to do with the 'research' that we set out to conduct). Indeed, we can develop a more speculative approach to the 'research event'. We can be more creative in how we enact a method assemblage in that we can explore how the 'research event' might unfold, and enquire into the various potentialities or virtualities latent in the event. In other words, following Isabelle Stengers (2010: 57), we can develop a method assemblage that 'affirms the possible..... actively resists the plausible and the probable targeted by approaches that claim to be neutral'.

Post-ANT and speculative design

It should be clear that there are numerous resonances between post-ANT and speculative design as portrayed in the foregoing. Both are invested in a sense of the openness of events, and both adhere to a practice of enabling that openness to unfold (for an alternative, see Latour 2008). If post-ANT (inspired by various related writings) has evolved a conceptual vocabulary for addressing the methodological processes of engaging with the 'possible' of an event, speculative design has developed a series of techniques (such as cultural probes) which have practically enabled entrée into the 'possible'. Needless to say, things are not so simple. As we have described anecdotally in this volume, post-ANT (or process-oriented social science) and speculative design (in the sense in which it is taken up in ECDC) have informed one another: in other words, in the extended event of their coming together, they have mutually changed. Thus, even within our own team, designers have become, albeit all too modestly, practised with using the concepts of post-ANT, and sociologists have become more comfortable with doing research through practice, through the process of practically working out how things might work (technically, materially, and aesthetically, as well as socially).

In this section, we discuss how the Energy Babble reflects the concrescence of post-ANT and speculative design. As a mixture or hybrid of the practical and the theoretical – a conceptual object and a material idea – it has been a key component in our method assemblage (which included site visits and probe exercises – see descriptions of what we did elsewhere in his volume). In particular, its specific design served to shape the possibilities that emerged in the research event of ECDC. These possibilities are at least twofold: they relate to what was emergent for the participants, but also what was emergent for us as researchers. The ambiguity and playfulness of the Energy Babble's operation – the ways in which its broadcasts varied in their 'sensibleness' – kept the research event 'open' (though as we see elsewhere in this volume, there is also a temptation amongst some participants to 'close down' its meanings). How did all this happen?

From the perspective of speculative design, we might say that the slow and meticulous process of practically developing the Energy Babble installed a capacity for it to surprise both designers and users. This is in part because the Babble's broadcasts do not make too much, or immediate, sense. The Babble thereby taps into the uncertain and emergent character of events (as understood in the present account). In the specific setting of energy communities and energy-demand reduction, the Babble potentially incites, not least in us as design researchers, 'interesting problem-making' around what counts as a community, what comprises information, and what can be understood by energy.

From the perspective of post-ANT, this practical 'keeping open' can be theorised through the figure of the 'idiot'. According to Stengers (2005a), the idiot 'resists the

and fully participating in a world that is richly interconnected and continually emerging.

consensual way in which the situation is presented and in which emergencies mobilise thought or action' (ibid.: 994). This is due to the proposition that the idiot does not make much sense when regarded in relation to the consensual or standard ways of framing an event. Because it resists easy comprehension, 'the idiot demands that we slow down, that we don't consider ourselves authorized to believe we possess the meaning of what we know' (ibid.: 995). As social researchers, our job becomes one of 'bestow(ing) efficacy upon the murmurings of the idiot, the "there is something more important" that is so easy to forget because it "cannot be taken into account", because the idiot neither objects nor proposes anything that "counts"' (ibid.: 1001). To be serious about the idiocy of the Babble is to be open about the openness of the research event of which it is a part – to use it for 'interesting problem-making' around what counts as a community, what comprises information, what can be understood by energy.

Concluding remarks

While we have, for ease of exposition, disaggregated the perspectives of speculative design and post-ANT in the foregoing, it should be apparent from the general discussion that things are far messier. There has been a mutual shaping – a co-becoming even – between speculative design and post-ANT in which each has assimilated elements of the other – respectively a certain vocabulary and a certain practical playfulness.

In terms of recent discussions about interdisciplinarity, we might say that the collaboration between speculative design and post-ANT has been 'ontological' (e.g. Barry et al. 2008; Born and Barry 2010) insofar as the concrescence of the various resources that each has brought into the research event has generated a new object of study – 'the possible' (of energy-demand reduction in the present case). However, this has not been without tension – tension that remains. Here, Stengers' (2005b) notion of an ecology of practices proves useful. For all the mutual shaping, designers and social scientists are nevertheless still oriented towards divergent, as well as common, intellectual communities. Each has a particular understanding of what constitutes energy-demand reduction that is informed by the particularities of their home discipline. So, there is inherent in this process of collaboration a strain born of connections to specific disciplinary ways of doing and knowing. And yet, given that such collaboration is also an event of mutual change, it serves as an occasion for reconfiguring the disciplinary terrain – of, so to speak, opening up the possibilities of new ways of working together.

References

Akrich, M., 'The De-scription of Technical Objects', in W. E. Bijker, and J. Law, eds, *Shaping Technology/Building Society* (Cambridge, MA: MIT Press, 1992), pp. 205-224.

Barry, A., G. Born, and G. Weszkalnys, 'Logics of Interdisciplinarity', *Economy and Society*, 37 (2008): 20-49.

Bijker, W. E., *Of Bicycles, Bakelites, and Bulbs: Toward a Theory of Sociotechnical Change* (Cambridge, MA: MIT Press, 1995).

Binder, T., E. Brandt, P. Ehn, and J. Halse, 'Democratic Design Experiments: Between Parliament and Laboratory', *CoDesign*, 11(3-4) (2015): 152-65.

Binder, T., G. De Michelis, P. Ehn, G. Jacucci, P. Linde, and I. Wagner, *Design Things* (Cambridge: MIT Press, 2011).

Boehner, K., W. Gaver, and A. Boucher, 'Probes' in C. Lury and N. Wakeford, eds, *Inventive Methods: The Happening of the Social* (London: Routledge, 2012), pp. 185-201.

Born, G., and A. Barry, 'Art-Science: From Public Understanding to Public Experiment', *Journal of Cultural Economy* 3(1) (2010): 103-19.

Brown, N., and M. Michael, 'A Sociology of Expectations: Retrospecting Prospects and Prospecting Retrospects', *Technology Analysis and Strategic Management*, 15(1) (2003): 3-18.

Callon, M., 'The Sociology of an Actor-Network: The case of the Electric Vehicle' in M. Callon, J. Law, and A. Rip, eds, *Mapping the Dynamics of Science and Technology* (1986): 19-34.

Cartwright, N., J. Cat, L. Fleck, and T. E. Uebel, *Otto Neurath: Philosophy Between Science and Politics*, (Cambridge: Cambridge University Press, 2008).

Connolly, W.E., *A World of Becoming* (Durham, NC: Duke University Press, 2011).

Deleuze, G., and F. Guattari, *What Is Philosophy?* (London; New York: Verso, 2011).

DiSalvo, C., *Adversarial Design* (Cambridge, MA: MIT Press, 2012).

Dunne, A., *Hertzian Tales: Electronic Products, Aesthetic Experience, and Critical Design* (Cambridge, MA: MIT Press, 2005).

Dunne, A., and F. Raby, *Design Noir: The Secret Life of Electronic Objects* (London and Basel: August/Birkhauser, 2001).

— *Speculative Everything: Design, Fiction and Social Dreaming* (Cambridge, MA: MIT Press, 2013).

Ehn, P., 'Participation in Interaction Design – Actors and Artefacts in Interaction', Paper Presented at the *Foundations of Interaction Design symposium*, Interaction Design Institute, Ivrea, Italy, 2003 <http://projectsfinal.interactionivrea.org/2004-2005/SYMPOSIUM%202005/communication%20material/Participation%20in%20Interaction%20Design_Ehn.pdf> [accessed 31 August 2013].

In short, the Babble sought to throw the communities' world of policy, technology, and

Fraser, M., 'Facts, Ethics and Event' in C. Bruun Jensen, and K. Rödje, eds, *Deleuzian Intersections in Science, Technology and Anthropology* (New York: Berghahn Press, 2010), pp. 57-82.

Gaver, W., 'Making Spaces: How Design Workbooks Work' in *Proceedings of CHI 2011*, pp. 1551–60.

Gaver W., A. Boucher, A. Law, S. Pennington, J. Bowers, J. Beaver, J. Humble, et al., 'Threshold Devices: Looking Out from the Home, *Proceedings of the 26th Annual SIGCHI Conference on Human Factors in Computing Systems,* Florence, Italy (New York: ACM Press, 2008): 1429-38.

Gaver, W., J. Bowers, T. Kerridge, A. Boucher, and N. Jarvis, 'Anatomy of a Failure: How We Knew when our Design Went Wrong, and What We Learned from it', Paper Presented at *Conference on Human Factors in Computing Systems*, Boston, MA, 2009.

Gaver, W., M. Michael, T. Kerridge, A. Wilkie, A. Boucher, L. Ovalle, and M. Plummer-Fernandez, 'Energy Babble: Mixing Environmentally-Oriented Internet Content to Engage Community Groups', in *Proceedings of the 33rd Annual ACM Conference on Human Factors in Computing Systems* (2015): 1115-24.

Gaver, W., P. Sengers, T. Kerridge, J. Kaye, and J. Bowers, 'Enhancing Ubiquitous Computing with User Interpretation: Field Testing the Home Health Horoscope' in *Proceedings of CHI ACM 2007* (2007): 537-46.

Hawkins, H., 'Geography, and Art, an Expanding Field: Site, the Body and Practice', *Progress in Human Geography* 37(1) (2013): 52-71.

Lang, P., and W. Menking, Superstudio: *Life without Objects* (Milan: Skira Milano, 2003).

Latour, B., *Science in Action: How to Follow Engineers in Society* (Milton Keynes: Open University Press, 1987).

— *Reassembling the Social* (Oxford: Oxford University Press, 2005).

— 'A Cautious Prometheus? A Few Steps Toward a Philosophy of Design (with Special Attention to Peter Sloterdijk)', Paper Presented at *Networks of Design* meeting of Design History Society Falmouth, Cornwall, 2008 <http://www.bruno-latour.fr/articles/article/112-DESIGN-CORNWALL.pdf>

Law, J., 'Technology and Heterogeneous Engineering: The Case of Portuguese Expansion', in W. E. Bijker, T. P. Hughes, and T. Pinch, eds, *Social Construction of Technological Systems* (Cambridge, MA: MIT Press, 1987), pp. 111-34.

Law, J. *After Method: Mess in Social Science Research* (London: Routledge, 2004).

Marres, N., and J. Lezaun, 'Materials and Devices of the Public: An Introduction', *Economy and Society*, 40(4) (2011): 489-509.

Michael, M. Anecdote in C. Lury, and N. Wakeford, eds, *Inventive Methods: The Happening of the Social* (London: Routledge, 2012a), pp. 25-35.

— '"What Are We Busy Doing?" Engaging the Idiot', *Science, Technology & Human Values*, 37(5) (2012b): 528-54.

— *Actor-Network Theory: Trials, Trails and Translations* (London: Sage, 2016).

Michael, M., and W. Gaver, 'Home Beyond Home: Dwelling with Threshold Devices', *Space and Culture*, 12 (2009): 359-70.

Mol, A., *The Body Multiple: Ontology in Medical Practice* (Durham, NC: Duke University Press, 2002).

Papanek, V., *Design for the Real World*, 2nd edn (London: Thames and Hudson, 1984).

Pignarre, P., and I. Stengers, *Capitalist Sorcery: Breaking the Spell.* (Palgrave Macmillan, 2011).

Stengers, I., 'The Cosmopolitical Proposal' in B. Latour, and P. Webel, eds, *Making Things Public* (Cambridge, MA: MIT Press, 2005a), pp. 994-1003.

— 'Introductory Notes on an Ecology of Practices', *Cultural Studies Review*, 11 (2005b): 183-96.

— *Cosmopolitics I* (Minneapolis, MN: University of Minnesota Press, 2010).

Storni, C., 'Unpacking Design Practices: The Notion of Thing in the Making of Artifacts', *Science, Technology, & Human Values*, 37(1) (2012): 88-123. Strengers, Y., *Smart Energy Technologies in Everyday Life: Smart Utopia?* (Basingstoke: Palgrave MacMillan, 2013).

Suchman, L. A., *Plans and Situated Actions: The Problem of Human-Machine Communication* (Cambridge: Cambridge University Press, 1987).

Whitehead, A. N., *Process and Reality* (Cambridge: Cambridge University Press, 1929).

Wilkie, A., and M. Ward, 'Made in Criticalland: Designing Matters of Concern', in J. Glynne et al., eds, *Networks of Design: Proceedings of the 2008 Annual International Conference of the Design History Society* (UK): 118-23.

Wilkie, A., 'Prototyping as Event: Designing the Future of Obesity', *Journal of Cultural Economy* (2013): 1-17.

Wilkie, A., M. Michael, and M. Plummer-Fernandez, 'Speculative Method and Twitter: Bots, Energy and Three Conceptual Characters', *The Sociological Review*, 63(1) (2015): 79-101.

Woolgar, S., and J. Lezaun, 'The Wrong Bin Bag: A Turn to Ontology in Science and Technology Studies?', *Social Studies of Science*, 43(3) (2013): 321-40.

practices into disarray, to see what new questions and possibilities might emerge. No

FEBUBABBLE

With 30 Energy Babbles on the loose, the month of February is all about turning on and joining in. Use your Babble to tell us what you are up to, how local projects are going, what's coming up and what has captured your attention in the news.

February will be a focus for our research activity, and with your contributions Babble will become a richer audio landscape.

If your Babble is still in a box, set it free and turn it on. If you would like to use Babble at an event, need help with wifi, or want to pass it to a friend, email info@ecdc.ac.uk or call 020 7078 5185.

Find out more at www.ecdc.ac.uk

Babble Gallery
Where does your Babble live? Send us a picture!

Jingle Jam
Compose a jingle to identify your community.

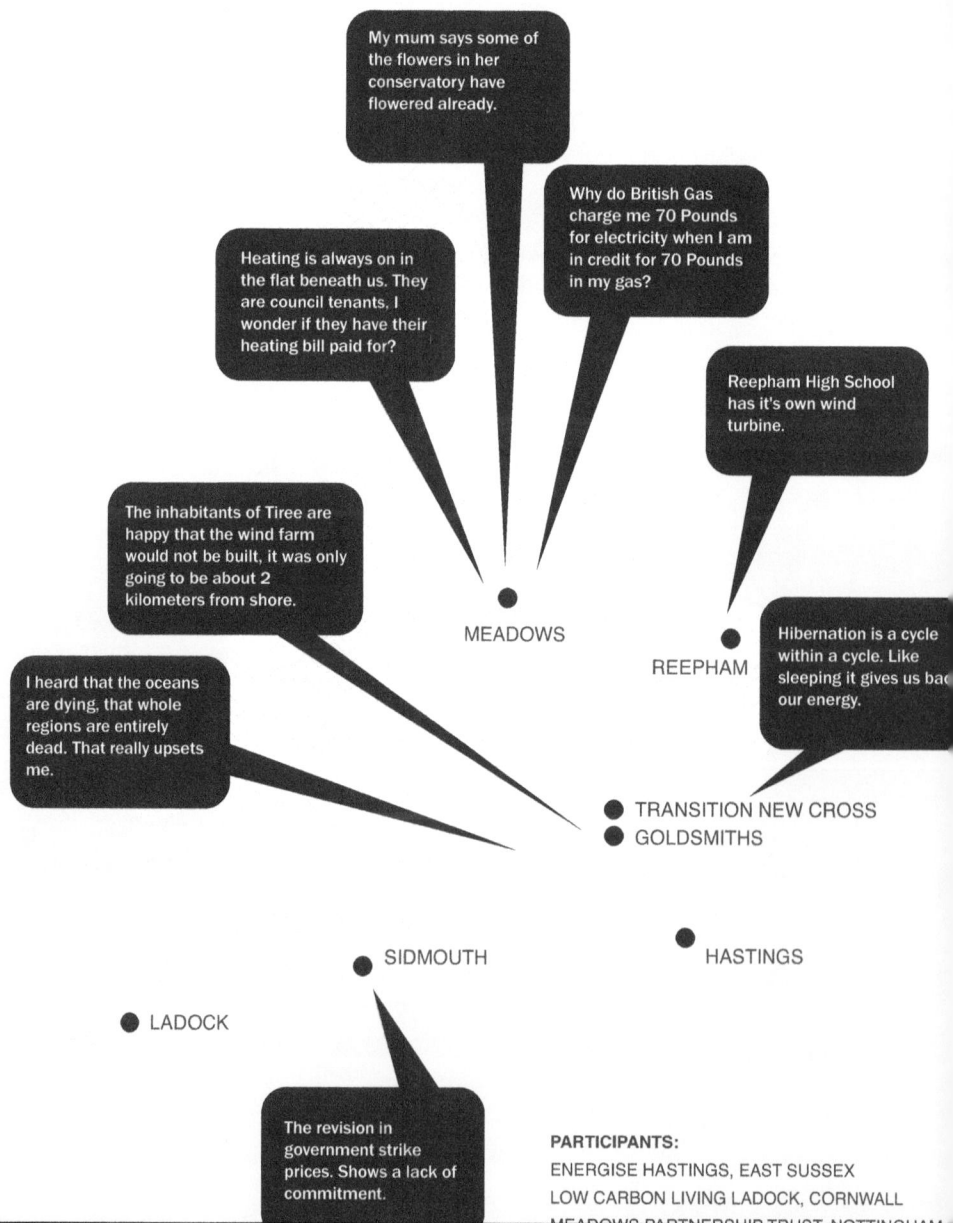

Newsletters were sent to the communities during the deployment

My mum says some of the flowers in her conservatory have flowered already.

Why do British Gas charge me 70 Pounds for electricity when I am in credit for 70 Pounds in my gas?

Heating is always on in the flat beneath us. They are council tenants, I wonder if they have their heating bill paid for?

Reepham High School has it's own wind turbine.

The inhabitants of Tiree are happy that the wind farm would not be built, it was only going to be about 2 kilometers from shore.

MEADOWS

Hibernation is a cycle within a cycle. Like sleeping it gives us back our energy.

REEPHAM

I heard that the oceans are dying, that whole regions are entirely dead. That really upsets me.

TRANSITION NEW CROSS
GOLDSMITHS

SIDMOUTH

HASTINGS

LADOCK

The revision in government strike prices. Shows a lack of commitment.

PARTICIPANTS:

ENERGISE HASTINGS, EAST SUSSEX
LOW CARBON LIVING LADOCK, CORNWALL
MEADOWS PARTNERSHIP TRUST, NOTTINGHAM

wonder, then, that people weren't quite sure what to make of the devices – depending o

Reactions from participants when presenting the devices

'How does this improve the social operational well being of the people who use it? .... If I make an investment how do get a payback?'

Goldsmiths. Would it be ok if we drop by at 11.45 to hand y̶o̶u̶ a Babble? It ̶t̶a̶k̶e̶s̶ ̶u̶s̶ about 30 ̶t̶o̶ set it up. See you later! Best, Liliana

Yes. That's perfect time ̶a̶ctually. :)

'It's a new type of thinking, you don't know what you'll get, it might just be chaos'

Great ! We will see you i̶n̶ bit!

'Devices for energy reduction are getting more and more complex and people's understanding hasn't. Some people cannot figure out a thermostat.'

'That is a very powerful sales tool'

Th̶

It's amazi̶n̶g̶. I love it so much already. The messaging reminds me of the barbed wire telephone system in wild west. Seriously- google i̶t̶ Thanks guy̶s̶

'If you put garbage in it, you will get garbage'

their interests, they might think about broadcasting the energy they'd generated or saved,

**Collecting the devices
after deployment**

or using it to recruit new members, or improving their social relations, or even just

enjoying it for the unusual device it was. Meanwhile, they were going about their work

Community as a state of mind
Liliana Ovalle

We picked up the last Energy Babble in deployment in July 2014, nine months after the trial had begun. The last device had been given to Jessica, one of our participants from Transition New Cross (TNX), a low-carbon community in South London. As in many of our deployment visits, we were welcomed with a cup of tea followed by a conversation about the experiences with Babble and a catch-up on the group's activities.

From our initial contact to the deployment of the devices, our visits to the communities provided us with a glimpse into the struggles and achievements of each group. In our last meeting with Jessica, we asked about the recent activities of the TNX group. 'Transition is a state of mind, it is more of an attitude,' she explained while telling us that the group had not met in the last six months. This answer contrasted with our first encounter with the group.

We met the TNX group when we attended one of their regular meetings at the Green Shoots Community Garden. At that time, we were at the stage of recruiting communities to participate in the project, and we were interested in working with a group local to Goldsmiths University.

Most the groups we had engaged with so far were communities that had received funding from the Low Carbon Community Challenge (LCCC), a DECC programme that provided funding for carbon reduction initiatives. While each group responded to different social contexts, this circumstance made them share particular characteristics: not only did these groups have the skills and human capital required to put together an application of that calibre, but they had also configured themselves into organised entities in order to be eligible for funding, from development trusts to registered energy service companies. In this context, TNX was a relatively emergent community.

At the time of our first meeting, TNX consisted of a group of local volunteers engaged in practices to reinvigorate the community and promote skills towards a sustainable future. The group had adopted the Transition model, a community model founded by Rob Hopkins in 2006 that provides guidelines for self-organisation towards resilience and reduction of carbon emissions. Mixed in age and backgrounds, the group was looking into initiatives that would allow them to establish themselves

and expand their activities, which so far included organising film screenings, up-cycling workshops, and gardening training. The ECDC project was an interesting opportunity for them to interact with other groups and to explore different approaches to energy reduction.

Throughout the beginning of the ECDC project, TNX was an active and engaged participant. This was particularly evident at the initial workshop at the Geffrye Museum, an engagement event that brought together members of the different communities participating in the project. With seven representatives, TNX had a strong presence. Its members actively responded to Cultural Probes, and participated in discussions on the ECDC blog.

During the development of the Babble, contact with the communities became more sporadic. We would send occasional updates via a newsletter about our process to touch base. As the project developed, contact with TNX became less tangible. By the time of beginning the preparations for deployment, the website had been inactive for over a year, and emails to the official email address were not responded to. After chasing personal

contacts, we learned that the group's communication platform had moved to Facebook. I posted an invitation to the deployment on their page and two of the members, Carlos and Jessica, decided to take part.

During visits, we learned that the group had not met as TNX for some time. One of the leaders of the group, who had driven most of the group's activities, had moved out of the area. Even though many of the members of the group still met and took part in activities in the local community garden, the consolidation of the group in the Transition Town format appeared to have dissolved.

The length of the ECDC project gave us the opportunity to observe how each of the seven participating communities had evolved and responded to different challenges, from coping with policy changes to fully implementing new technologies. The case of Transition New Cross highlighted some of the struggles that communities can encounter. In the space of three years, the group had gone from being a collective effort to being an individual state of mind, detached from action.

This example illustrates how essential factors such as the cohesion and management of a group often rely on a few individuals, and it takes a change of circumstances to dissolve the bonding capital required to build a group. From a leader who moves address, to the lack of time for volunteering, the consolidation of a community of practice can be a precarious process. In this panorama, public funding can play a powerful role for communities to shape, as it can act as a catalyst to accelerate the process. But, equally, it can bring specific obligations and demands, configuring the communities in specific ways. The groups that are left outside the umbrella of policy remain susceptible to disintegration unless they can meet the right conditions.

Engaging the community

As we travelled the country to different communities, we heard a myriad of reactions from the community participants to the Babbles. For most of the people we talked to, the Babbles were unsettling, giving rise to ambivalence and confusion. The bricolage of text and music they offered was sometimes interesting, sometimes amusing, but rarely of real practical use. Nevertheless, the deliberation and care in the Babble's design and the technical achievement in its production was evident. This was clearly no simple failure. The puzzle presented by the Babble, and the complicated ways it fitted and misfitted participants' circumstances spurred long conversations. Soon after the field trial was over, we reported the Babble project to a preeminent conference in Human Computer Interaction (HCI). How to explain what we had found to an audience of computer, behavioural, and social scientists, with only a subset of members versed in design and the humanities? This extract from our paper simultaneously reports details of what our participants told us, and how we tried to explain this to ourselves and to our audience.

Opposite, excerpt from: William Gaver, Mike Michael, Tobie Kerridge, Alex Wilkie, Andy Boucher, Liliana Ovalle, and Matthew Plummer-Fernandez. 2015. Energy Babble: Mixing Environmentally-Oriented Internet Content to Engage Community Groups. In Proceedings of the 33rd Annual ACM Conference on Human Factors in Computing Systems (CHI '15). ACM, New York, NY, USA, 1115-1124.

about their experiences with them, their responses addressed many issues. Bottom line:

Sustainability & Recycling

items, ask questions and report energy use, all interspersed with occasional musical interludes and lapses into nonsense. The majority of content is related to energy and the environment and thus the devices present themselves as strongly focused on sustainability, though a fair amount of 'off-topic' content also creeps in from Twitter™, from following links, or from participant inputs. Our question was how our participants, all committed to environmental concerns themselves, would engage the Energy Babble.

LIVING WITH THE ENERGY BABBLE

We deployed a total of 21 Energy Babbles to members of the communities in a series of meetings at their locations (Figure 5). Each community received 3 or 4 devices, which were usually given to volunteers present at the meetings, though in a few cases extras were left for later distribution. The remaining 5 Babbles were distributed to team members, with 2 going to people more loosely connected to the project. Volunteers lived with the Babble for varying periods averaging about six months.

In the rest of this section, we briefly describe what our participants told us about their experiences with the Babbles. The majority of reports come from discussions when we deployed the devices, or several months later when we revisited the communities to pick up the devices. Others come from documentaries by an independent filmmaker hired to help us assess the field trial.

Initial Expectations and Impressions

We packaged the Energy Babbles, associated documents and equipment in custom-made cardboard boxes for transport. During deployment events, these were usually positioned visibly, but unopened, during an introductory presentation in which we reviewed the project. Then we would unpack a Babble device and describe how it worked. Because it took some time to set up one of the devices for demonstration, during these initial descriptions the group had not yet heard the system. Typically, then, initial comments and questions revealed a mix of assumptions, expectations and responses to the devices.

Initially, many participants expected us to produce a tool that would directly help them reduce energy consumption — or as G, from the Meadows, put it: *'We thought we were going to get a gizmo to save energy'*. When it became clear the Babble did not serve this purpose, they looked for other utilitarian pay-offs. In Hastings, for instance, an engineer asked *'How does this improve the social operational wellbeing of the people who use it? If I make an investment how do get a payback?'* and explained *'I wanted it to solve a problem'*. These discussions tended to encourage the news/communication interpretation of the device, which mollified many skeptics. For instance, the engineer realised it could be used to broadcast the energy output of their renewables: *'Babble could bring this information to people'*. Similarly, in Laddock a group member championed the Babble for using the British Raspberry Pi technology, and because he saw potential for it to broadcast his car battery operated DIY domestic electricity system.

Building on this, in several of the groups volunteers saw potential value in the Babbles as a kind of marketing tool for promoting their groups and environmental concerns more generally. The Hastings engineer, for instance, described broadcasting energy generation figures as *'a very powerful sales tool'*. In Sidmouth, the group speculated about deploying the Babble in a local energy shop, or using it as a recruitment platform at an Alternative Energy Vehicle show. In Reepham, the group decided that one of the devices should be free to roam, initially to the Post Office and later to a variety of environmental events.

Some people were happier to relinquish a utilitarian interpretation of the Babbles during the deployment events. For instance, after listening to the device during the Meadows deployment, D decided that they would name their Babble 'Finnegan', in a reference to James Joyce's Finnegans Wake. She explained that this was because the output is like *'a stream of consciousness'*. In New Cross, J sent an SMS message after her Babble started working: *'It's amazing! I love it so much already. The messaging system reminds me of the barbed wire telephone system in Wild West. Seriously - Google it. Thanks guys. :)'*

Installation and Accommodation

Installing the Babbles involved configuring the devices to local router settings, dealing with security, and setting it up to communicate using the router's wireless network. In many instances this proved unproblematic, but in some cases, including deployment events, it proved more difficult. While none of the problems we encountered were insurmountable, they seemed to demonstrate to potential volunteers the possible inconveniences of borrowing a Babble. More serious problems arose with some of the devices we left behind. For instance, in Reepham problems with a local firewall prevented the Babble from being installed in a local primary school. Other devices were borrowed but never installed, possibly because of the perceived difficulty of set-up. Pragmatic issues were salient even for imagined deployments: for instance, in Sidmouth ideas for showing the device at the Alternative Energy Vehicle show involved thinking about powering it via a car with solar panels, and achieving mobile internet access.

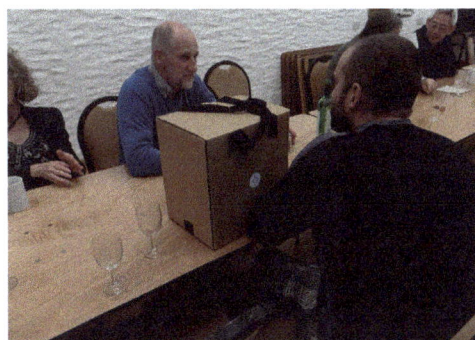

Figure 5. A deployment event.

nost admired the Babbles aesthetically, but not functionally. To explain this, they told

Visual and Auditory Aesthetics

The Babble has an idiosyncratic aesthetic that extends from its physical design to its auditory output. Most of the volunteers found this appealing. For instance, in Ladock, J said she and her husband appreciated the Babble because it was '*nice, funky looking thing*'. In the Meadows, P, an art tutor, said it was a '*really nice object*' and that the microphone was a '*lovely visual element*'. This appreciation was mixed with some bemusement, however. Several people remarked on it having a 'retro' appearance, or as J from New Cross put it, the Babble looked '*like my gran could have had one*'. It also was compared to kitchen appliances. For instance G from Ladock, told us '*it looked like a food processor... a bit quirky*', while G in the Meadows told us that visitors to his home usually asked jokingly why he had a blender in his living room. The glass elements, too, attracted a mixed reception: R from Reepham described the Babble as '*beautifully made in hand blown glass*', but its fragility was a worry for the librarian of a secondary school where it was installed.

Despite some initial concerns, we received no complaints about the synthesised voices used by the Babble, and several people remarked favourably about their clarity. On the other hand, the fact of it being an audio device could be disruptive. G, in Ladock, told us he had moved it from the kitchen to the living room because it was a '*conversation stopper*' for the family. L, in Reepham, the mother of a young baby, found it '*annoying and noisy... not really for a home*'. R, her husband, told us he had relocated it to his office because of this. In the Meadows, G said the Babble '*annoys everyone*' so he turned it on when he was alone.

The intermittent nature of the output could also be unsettling. '*A few times it frightened the living crap out of me!*', J in the Meadows told us. She elaborated that once when the office was completely silent at around 9pm, the Babble had given her a fright when it unexpectedly came on. She also complained that it '*didn't talk on cue*' when she showed it to visitors, and (like other volunteers) would have liked to be able to replay interesting outputs.

Babble as a Source of Information

Volunteers often oriented to the Babbles as a potential source of information. This is not surprising given that the audio was designed in the style of an automated news broadcast, that many of the volunteers showed a propensity to seek utilitarian explanations for the device, and that we tended to encourage these explanations to reassure them about the system, rather than foregrounding a description and defence of the Babble as a playful, reflective system.

By the end of the field trial, however, many volunteers expressed disappointment in the Babble as a source of environmental news. In Ladock, G told us that he did hear information '*which was interesting*', but explained that he did not follow many alternative sources of environmental news. In contrast, his colleague J told us that she welcomed the Babble as a source of new information, but '*disappointingly, not as much as I hoped*'. '*It seemed a bit sparse,*' she explained, and '*very repetitious*'. In Reepham, R subscribes to DEC emails that he looks at in the morning: '*if they're about something I'm interested in I read them*'. He told us that the Babble never provided relevant information of which he hadn't been aware.

A recurring theme in discussing the Babble was that too much of its output was irrelevant. J, in Ladock, for instance complained that there was '*a lot more of the jumbly stuff and less of the straight stuff*'. '*I tend to be on the serious side*', she explained, and '*definitely the balance was wrong*'. Considerations such as these led to suggestions for filtering the output. For instance, R in Reepham speculated that the Babble could be more like a radio: '*you might have one stream about transport, another about food, about heating the home...*', describing the result as '*far more relevant*'. To our suggestion that mixing streams might support serendipitous connections he was doubtful: '*people's attention spans are getting shorter*', he explained, so they would get bored before putting things together themselves. J, in the Meadows, also said that the Babble gathered too much irrelevant information and that it needed a '*filter*' to focus on reporting news about communities and government, '*rather than oil and gas*'. G, also from the Meadows, echoed this, suggesting the Babble could be an app with buttons to switch on and off channels of information—a '*filtering mechanism*'.

There was no clear consensus across volunteers about which streams of information were worth hearing, suggesting that the ability to select among them would lead to a more utilitarian design. In the Meadows, for instance, P found the energy reports frustrating as they didn't mean anything to him, while G, interviewed separately, said they were '*really really good*'. In New Cross, P reported that she couldn't follow the '*technical information*', referring both to the energy reports and the information on renewable systems. For her, '*you can connect more to personal comments, to the emotional side of energy*'

Babble as a Medium for Communication

Volunteers had mixed reaction to the ability to input and hear comments using the Babble's microphone and SMS facilities, and this was reflected in the relatively few messages they left on the system (about 35 over 5 months). There was an evident reluctance to enter messages. During the first weeks of living with Babble, for instance, J in New Cross made a few contributions using the microphone. Eventually her reaction became more of an '*internal conversation*'. She reported that when she reached for the microphone she felt nervous about saying something important to the system. Similarly, R from Reepham told us that he didn't input much because he has '*controversial views*' and didn't want to '*upset anyone*'. He recounted how he had heard something on the Babble that he disagreed with, but refrained from expressing his views because he considers them quite controversial. '*I was aware that DECC might be listening, I want to come across as quite conservative, you don't know who is listening*'. In the Meadows, G was concerned to prevent his

us in great detail about their struggles and successes dealing with regulations an

Sustainability & Recycling

daughter from saying silly things into the microphone (when asked what those might be he responded '*are there any fit hotties out there?*'). Nonetheless, many comments diverged from clear relevance for environmental concerns, and G was annoyed by messages he thought trivial.

The reluctance to contribute to the system ran counter to appreciation for the content that did appear. For example, J from the Meadows said she would have liked to hear more from the other communities, especially '*stories and tips on how they're dealing with these issues*'. In New Cross, J enjoyed the comments: '*you can connect more to personal comments, to the emotional side of energy*'. Conversely, G in the Meadows would have liked to read out his household energy use, while P would have liked to broadcast his solar energy production; however there was little reflection about who the audience for these figures might be.

Finally, we had some indications that the lack of user inputs into the Babble reflected a lack of interest in communicating with other groups more generally. R, in Reepham, was clearest about this: he told us that while he occasionally kept track of what other communities are doing, differing circumstances meant that '*what might not be right for them, might be right for us*'. He might check for good ideas but unless something was '*revolutionary*' there wasn't much use in this. Equally, he liked telling people what worked in Reepham, but described this as '*reactive not proactive*'—his group doesn't proselytise '*the way Transition Towns do*'.

Appreciation for the Babble

Despite the lack of clear success for the Babble as a utilitarian information or communications product, all the volunteers we spoke with were largely positive about it. In part this reflected appreciation for it as a well-finished, device that could fit the home (Figure 6). In part, it stemmed from admiration for the Babble as a novel technical device. In Reepham, for instance, R found '*stimulating*' the way it uses audio rather than visual/text as a way of encountering social media. In the Meadows, P speculated about extending the Babble's technology, for instance to automatically tweet about his solar panels, or to nag him about his bad energy habits.

Admiration for the Babble as a novel technical and aesthetic device blended with its value as something to show to other people. C from Hastings, for example, was effusive about the Babble, describing the novelty of the device and the attention it had garnered at work, where she originally installed the device, and at home, where she took it later. R from Reepham described it as '*a curiosity for visitors*' that he enjoyed to members of a number of other environmental organisations with whom he worked.

Finally, several volunteers expressed appreciation for the Babble as a source of ambient awareness of environmental action. In New Cross, J told us it was reassuring to hear evidence of expertise: '*Thank God for people who know the technical bits, it's strengthening to hear that there are people out there in charge*', and more generally that the

Figure 6. Babble in a volunteer's home

Babble gave her a sense of a larger community concerned with environmental issues: '*it makes you think that you are not alone in thinking about saving the world*'. In the Meadows, J expressed a more abiding affection for the presence of the Babble: '*aw, I'll miss him actually. It was nice to have him on in the background, I'm used to it now. Its quite aptly named, Babble*'.

Babble and Wider Conversations

The accounts above all reflect discussions centred fairly closely on the Babble system as a product. What became striking to us, however, was the way that our conversations with the volunteers frequently opened from an initial concern with the Babble's usability, functionality and aesthetics to encompass the broader and more particular issues, practices and controversies with which our volunteers were living. Though these discussions may be of questionable relevance for assessing the Babble as a *product*, we suggest that these conversations and the insights they revealed can be viewed as an outcome of the Babble as a *research tool*.

For instance, at the Meadows, during a suggestion that the Babble content should be filtered to focus on communities and government '*rather than oil and gas*', J suddenly exclaimed '*except that British Gas are bastards!*', and conversation with her diverged into lengthy complaints about DECC's lack of support, British Gas call centres, and pigeon droppings building up under solar panels. In Ladock, our conversation with J about the Babble soon expanded to include her complaints about the hurdles involved in securing government funding for environmental work ('*we think they're rubbish*'), and the frustrations of not being able to give away radiator backdrops, energy monitors and LED down lighters at an Energy Fair she organized ('*it was a total failure*' that '*didn't engage the people we set out to engage*').

Also in Ladock, G described their attempts to put up a new wind turbine that was rejected by the council '*on spurious grounds*'. He attributed this to '*about half a dozen*' residents who spread '*a lot of misinformation*' about how the Low Carbon Living group were out to '*line their own pockets*', culminating in '*a minor punch-up*'. Like J, he expressed

technologies within their communities. It seemed that Babble's carefully crafted

frustration at the difficulty in reaching out to dissenters within the community (there's '*no forum to talk to those people*'), and also with the government: '*lots of businesses are starting up then going to the wall because the government keeps changing the rules'*. These complaints were mixed with pride in the group's achievements. For instance, he referred to a '*story'* he put on the Babble about how on a sunny day he used his PV to charge his car and heat water: people were impressed that he could '*drive 75 miles and have hot water for absolutely nothing'*. He concluded that '*you can't depend on the government to do things, you can depend on the community to do things'*.

A notable theme that emerged from several volunteers had to do with the entanglement of energy concerns with other issues. For instance, R in Reepham told us he would be going to Buckingham Palace to be honoured for his contributions to energy efficiency, but said that he'd like to be recognised for the work he does that goes beyond that. The Babble should go beyond energy, he told us, to address fuel poverty, transport poverty, and take a '*holistic*' view. "*Energy is a key part of it but the stories are stories about many other things*' he said, '*It's too sterile if you look at only energy*'.

Similarly, in Ladock J told us that her husband refuses to be involved with Low Carbon Living because he sees their efforts as futile. '*We should be lobbying*' she said, and mentioned the social networking activist groups Avaaz and 38Degrees as effective ('*though I understand their limits*'). She also does work with Christian Aid. '*You tend to see how it all fits together - the international aspects of climate change*'. For instance, when Christian Aid pointed out that climate change harms the poorest first, she thought they were off-topic, but then realised it was true. This led her to realise that '*how we treat the world and how we treat other people, they're all linked*'.

DISCUSSION: UNDERSTANDING THE BABBLE

Taking seriously the idea that the Babble played an important role in sparking the intense discussions we had with our volunteers suggests that we move beyond assessing the system according to the utilitarian characterisation of it as an information and communication product. Turning to the reflective interpretation of the Babble instead, as a system that gathers and 'intensifies' the existing state of discourse around energy practices, may give us another perspective on how the system worked as a research tool in our meetings with the participants, by serving as an independent actor that helped shape conversations leading to better understandings of the communities and their concerns.

A simple version of this account would suggest that the Babble should be understood as a research tool that was successful, rather than simply as a utilitarian information/communication product that was less so. The distinction between these roles is not clear-cut, however. The Babble was never seen purely as a prototype product, either by the volunteers or ourselves: we never planned to

produce it commercially, and they were always aware of it as part of a research project. The Babble was never solely a research tool either: it was offered seriously for long-term use, and participants engaged with it not only to further their discussion with us but to engage with the material it offered in its own right. The product and research-tool faces of the Babble are interdependent. Here we discuss several conceptual handles on how this might be understood.

To start with, it is helpful to consider the Babble in terms of the conceptual character of the 'idiot', who, in Stengers' [13] account:

> resists the consensual way in which the situation is presented and in which emergencies mobilize thought or action. This is not because the presentation would be false or because emergencies are believed to be lies, but because "there is something more important". Don't ask him why, the idiot will neither reply nor discuss the issue.... the idiot demands that we slow down, that we don't consider ourselves authorized to believe we possess the meaning of what we know (p. 994)

From this point of view, the Babble can be seen to act as an idiot within the energy communities who used it (see [17] and Michael [9]), by confounding expectations of how technologies should contribute to the communities' work. This was evident both during the deployments, when the Babble surprised and confused volunteers who were expecting some sort of demand reduction meter, or at least a clearly utilitarian design ('*I wanted to solve a problem*'), and throughout the project, as volunteers struggled to make sense of what it was doing. Instead of acquiescing to 'the consensual way in which the situation is presented', the Babble implicitly suggested that in the confused flow of messages about energy use, policy shifts, new technologies, and seeming irrelevancies "there is something more important".

But what is that 'something that is more important'? The Babble never says, but given its output this might include keeping in touch with emerging policy, sharing best practice, being aware of energy sources and demand, and joining with other communities—the very concerns identified as important by the funding programme that supported the project. But the Babble does this in the most literal, even stupid, way, and the volunteers resist it. They counter by insisting that policies are ever-changing and wilfully made difficult, that what works for one community may not work for another, that it is difficult to find meaning in statistics about energy, and that there is limited value in further contact with other communities. From this point of view, the roles are reversed: it is the Babble that presents the 'consensual way in which the situation is presented', and the community volunteers who are cast as idiots, asking the Babble, and us, and the policy-makers, to slow down, because we do not 'possess the meaning of what we know.'

opacity elicited this deep reflection on their own situations. Later, once the field trial

Sustainability & Recycling

Our conversations with them at the end of the field trial, then, can be seen as reflecting their pent-up responses to the obduracy of the Babble. Yes, the Babble may be right in saying that there are larger concerns at play than can be addressed by energy demand meters, but what is needed is not simply more policy, more news, and more communications. On the contrary, they told us, we need better filtering, better ways to talk about energy, better situated ways to communicate, and recognition that energy use is situated in a wider landscape of local and global issues such as inequality and sustainability. And through this, they revealed their realities, helping us to understand that these 'communities' are shifting collections of people who constantly reconfigure themselves, and who do extraordinary work to negotiate changing policy opportunities and obstacles, to filter information about new technologies, to reach out within their own communities, and to understand when it is worth communicating more closely with others.

In the end, the Babble might be understood in terms of DiSalvo's [4] account of how design can play a part in constructing publics. Following Dewey, DiSalvo suggests that publics form around issues, and that design can participate in this by bringing issues to prominence. He suggests two primary tactics for this: *projection*, in which designs suggest possible future manifestations of current trends, and *tracing*, in which design is used to make clear the history of current situations. To this, the Babble might add a third tactic: *concentration*, in which current accounts and discourses about an issue—in this case energy—are brought together to form, not just a neutral representation, but a focused stream that inundates listeners with the many different and potentially incompatible ways that that issue is discussed, legislated for, measured and worried about.

From this perspective, the Babble might form 'a public' not just via the issues that comprise it, but the issues raised by the incoherence of the babble itself. Thus the Babble begins to point toward a public that emerges out of an oscillation between different local and collective communities, variously in competition and united, informed and frustrated. Moreover, the discussions occasioned by the Babble suggests that support for situated, local communities requires better appreciation of the morass that the publics/communities must negotiate. This includes the competition/lack of communication between communities, as well as the commonalities of being placed in a relation of competition by the structure of government project funding. In highlighting these issues, the Babble may also help (re)configure a public of HCI researchers, funders and policy makers to concern itself with these realities of energy communities rather than, simply, technologies focused directly on energy demand reduction.

ACKNOWLEDGMENTS

We are deeply grateful to the energy communities for their participation in this research.This project was supported by the Economic and Social Research Council's award no. ES/I007318/1 and by the European Research Council's Advanced Investigator Award no. 22652.

REFERENCES

1. Abrahamse, W., Steg, L., Vlek, C., & Rothengatter, T. (2005). A review of intervention studies aimed at household energy conservation. *Journal of environmental psychology*, *25*(3), 273-291.
2. Brynjarsdottir, H., Hakansson, M., Pierce, J., Baumer, E., DiSalvo, C., and Sengers, P. 2012. Sustainably unpersuaded. CHI 2012, 947-956.
3. Callon, M. (2004) The Role of Hybrid Communities and Socio-Technical Arrangements in the Participatory Design. Journal of the Center for Information Studies 5(3): 3–10.
4. DiSalvo, C. (2009). Design and the Construction of Publics. Design Issues, *25*(1), 48-63.
5. Dourish, P. 2010. HCI and environmental sustainability. DIS 2010, 1-10.
6. Gaver, W. 2009. Designing for Homo Ludens, Still. In (Re)searching the Digital Bauhaus. Binder, T., Löwgren, J., and Malmborg, L. (eds.). London: Springer, pp. 163-178.
7. Gaver, W. 2011. Making spaces. CHI'11, 1551-1560.
8. Kerridge, T. (Forthcoming). Designing Debate: The Entanglement of Speculative Design and Upstream Public Engagement with Science and Technology. PhD thesis, Goldsmiths, University of London, London.
9. Michael, M. (2012). "What Are We Busy Doing?" Engaging the Idiot. Science, Technology & Human Values, 37(5), 528-554.
10. Schultz, P. W., Nolan, J. M., Cialdini, R. B., Goldstein, N. J., & Griskevicius, V. (2007). The constructive, destructive, and reconstructive power of social norms, Psychological Science, 18(5), 429–434.
11. Sengers, P. and Gaver, W. (2006). Staying Open to interpretation. Proc. DIS 2006.
12. Shove, E. (2004). Comfort, cleanliness and convenience. Berg.
13. Stengers, I. (2005). The cosmopolitical proposal. Making things public, 994-1003.
14. Strengers, Y. (2013) Smart Energy Technologies in Everyday Life: Smart Utopia? Palgrave MacMillan.
15. UK Research Council (2010), Energy and Communities call for proposals. http://web.archive.org/web/20100813103416/http://www.esrc.ac.uk/ESRCInfoCentre/opportunities/current_funding_opportunities/Energy_and_Communities_Collaborative_Venture.aspx (accessed 22.09.14).
16. Wilkie, A. (2013) Prototyping as Event: Designing the Future of Obesity, Journal of Cultural Economy: 1–17.
17. Wilkie, A., Michael, M., & Plummer-Fernandez, M. (2014). Speculative method and Twitter. The Sociological Review.
18. Wilkie, A., Michael, M., Kerridge, T., Gaver, W., Ovalle, L,. DiSalvo, C., Gabrys, J., (2012). Design, STS and cosmopolitics: From intervention to emergence in participation and sustainability. Proc. EASST 2012.

Exhibiting and engaging

After the field trial, the Babbles changed their role from being devices that lived, however awkwardly, in the communities, to being devices for communicating our research to a variety of new publics. They were shown in exhibitions organised by Karen Henwood's group from Cardiff University and ourselves, which displayed various outputs from projects in the overarching Energy Programme that funded them. This reached an audience including hipsters from Shoreditch, schoolchildren, and members of the Welsh National Assembly. Over the subsequent years, Energy Babbles have also appeared in several Australian exhibitions and are scheduled to appear in the US. In these settings, they are presented as representing cutting-edge interaction design as well as new approaches to designing for environmental engagement.

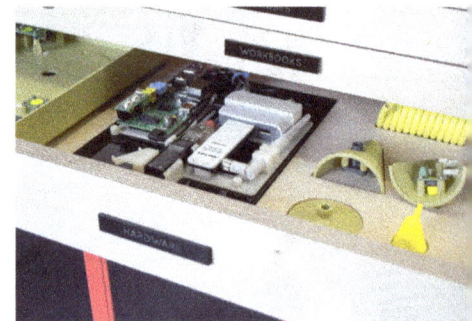

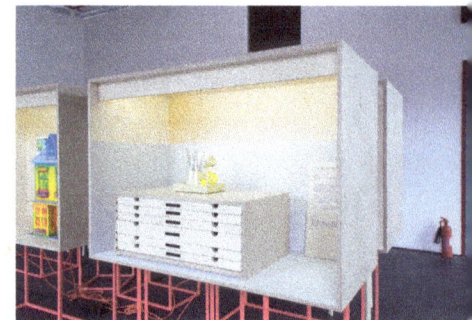

We joined Karen Henwood's group from Cardiff University to stage 'A Sense of Energy', an exhibition of visual data produced by the projects funded by the RCUK Enegy programme. The first showing was held at the White Building in Shoreditch, in the midst of many of London's creative industries.

groups to stage an exhibition of our work, where publics from school children to nationa

'A Sense of Energy' was staged a second time at the Senedd in Cardiff, home to the National Assembly of Wales, accompanied by onsite workshops and lectures. This exposed the work to a very different set of publics from those at the White Building, including assembly members and schoolchildren.

politicians encountered the Babble alongside other energy-related research. Other curators

Three years of living with an Energy Babble

Katherine Moline

When I contacted the Interaction Research Studio with an invitation to exhibit Energy Co-Designing Communities (ECDC) in Sydney, I was drawn to it as an exemplar of the impactful research in art and design that was developing in university research centres (UNSW Galleries, 2014). With a very short lead time, my curatorial aim for the exhibition 'Feral Experimental: New Design Thinking' was to test the limits of possibility in an exploration of the boundaries and intersections of experimental practice in design thinking, speculative design, participatory design, and co-design. Thankfully, Bill Gaver, Tobie Kerridge, and the Interaction Research Studio said yes. The project's initial attraction was its upscaling of cultural probes across the UK to address the depletion of energy resources with lay experts in low-energy communities. During hours of hunting through the project material I came to appreciate the depth of engagement, the open-source nature of ECDC's process, and the collected data that the studio made freely available online. It provoked me to think carefully about how the nuanced debates about and between design methodologies as new specialisations opened up or ghettoised design. In bringing together real-world applications, my hope was that 'Feral Experimental' would make public how contemporary art and design addressed significant chall-enges with new hybrid approaches that were not possible without an interdisciplinary agenda.

We opened the package that arrived from the UK. I was shocked when I realised the package contained the Energy Babble. I was shocked because the Babble had expanded in my imagination to two or three times the size of the device in front of me. Once extracted from its custom package, the gallery preparators, technicians, and I searched the Babble's surface for clues as to how to make it work. After a long discussion on gallery concerns that a custom electronic might burn down the newly built gallery, I snuck it down to the resident tech wizard in the university's computer services. He delicately unscrewed the moulded plastic casing so that we could see what was inside. Not easily impressed, his first comment was on the elegance of the object's internal layout and he carefully explained how well thought out the design was. In a volley of rapid-fire emails, the Babble became increasingly opaque to me as its innards were described in detailed technical language. It slowly dawned on me that we couldn't play it live in the exhibition because of gallery regulations and the fact that Australia was asleep and the gallery was closed when the users of Babbles in the UK were broadcasting live. More than once I wondered if the Babble was designed to confound us, not only practically but also conceptually. Was the Energy Babble a Surrealist game about exhibiting a design named speculative but that functioned as a co-design. I was in conversation with a gallery visitor who was studying it closely one day, and she exclaimed that it was a design named co-design but that functioned as speculative in that it formed a large collective that could broadcast and share ideas and debate the ethics of using energy indiscriminately (petrol, electricity,

solar power). We debated whether the Babble was a Surrealist inversion about autonomy and the emotional viability of living ethically. In that conversation, we contemplated whether its function as a CB radio was to avoid the Orwellian connotations of Big Brother surveillance on the internet.

During an early gallery tour of the exhibition, further conversations focused on the Babble's functional resemblance to a CB radio. Once discussion turned to its purpose to connect special interest groups in the UK, questions were raised about whether it was engaged with the Internet of Things, or an experimental device that quite literally demonstrated the importance of small details in its conceptual framework, and whether its elaborate processes that engaged end-users as participants, and its elegantly retro execution, were important for designing interaction in a Human-Computer Interface for low-energy communities.

The Energy Babble's plain, domesticated appearance that juxtaposes a conical funnel with a coiled phone cord (phased out in the 1990s) sitting atop a phone charger box prompted one student to write about the design as an innocuous kitchen appliance that was 'disruptive' because if 'left in a public space, the microphone may pick up some irrelevant and absurd input'. The Surrealist ambiguity and Orwellian undertones about what it was doing in the exhibition prompted colleagues to debate its intentions (a surprisingly frequent art school concern about the ethics of design). Another student pointed out that people just get annoyed with devices, citing mobile

became interested and the Babbles were exhibited in several shows in Australia and wil

phones as an example, but she saw it as 'a great tool for rethinking energy issues that were raised by government policies, local commercial activities, individual efforts, and small communities'.

Over three subsequent exhibitions that developed from 'Feral Experimental', I led curatorial teams who also saw the relevance of ECDC to the communities in which the exhibitions took place. Whereas in Sydney 'Feral Experimental' focused on leading-edge examples of art and design that addressed contemporary challenges, in Melbourne with co-curators Laurene Vaughan and Brad Haylock the exhibition 'Experimental Practice: Provocations In and Out of Design' brought together works that raised important questions about design, data and impurity (RMIT Design Hub, 2015). In the catalogue essay, I described the 'exhibition program to which 'Experimental Practice' contributed' as aiming 'to modify the exhibition and symposium/ workshop agenda in each site according to local knowledge, and according to the ongoing development of selected works over time', and cited the Energy Babble and ECDC as a demonstration of 'how large-scale projects evolve' when design is exhibited as works in progress rather than fixed or finished in a gallery context. I contended that, unlike the everyday understanding of design as a form of expertise based on control, the exhibition, workshops, and panel discussion extended the debate beyond the stated intentions of the practitioners (Moline 2015b: 8). In the case of the Babble, however, the studio's intentions were ambiguous in that they aimed to engage playfully with irrational aspects of domestic energy reduction (Gaver, Michael, Kerridge et al. 2015: 1118). As I discuss shortly, this ambiguity meant that gallery visitors suggested wide-ranging interpretations of the design. Within the remit of the Studio's intentions, these interpretations negotiated contradictions between multiple factors, including consumption

practices and environmental aims, that must be rethought for design, and indeed art, to make a difference to the contemporary challenge of climate change.

In Brisbane, 'Experimental Thinking/Design Practices' aimed to emphasise embodied knowledge and further complicate questions about design in terms of curatorial and research practice, as well as making (Griffith University Art Gallery, 2015). Co-curated with Peter Hall and Beck Davis, this exhibition explored how lived experience informed and inspired design. The catalogue essay described the unifying aim of the exhibition series that was generated from 'Feral Experimental' as drawing together a number of approaches to the challenges of global warming, big data, and embodied experience in the digital context. I explained that the exhibitions aimed to search for the holes in the fence that dingoes, a species of dog unique to Australia, are expert at sniffing out. 'Through these openings', I proposed that 'connections between categories are made and disciplines communicate with each other to develop new approaches for addressing contemporary concerns' (Moline 2015c) In the catalogue essay I spelt out that 'crossovers between co-design and speculative design suggest that such categories are not clear-cut. Recent co-design has engaged imaginatively with alternatives to the status quo via the cultural probes of speculative design, while speculative design that engages communities in reimagining the future has been developed with co-design's focus on lived experiences. In other words, in practice, both approaches combine strategies to imagine possible futures with a greater number of stakeholders, and recombine technologies, to address wide-ranging issues' (Moline 2015a) In greater depth, here I explained the competing definitions of speculative design, and that one of the contributors to ECDC, Mike Michael, defined it as a framework for engaging the public in science and

technology studies, which explains what he sees as 'overspills' and public responses that exceed the parameters established by researchers. Rather than ignore or 'sanitise' unexpected events, Michael sees them as a source for insights that generate new approaches to design (Michael 2012) Showing ECDC in Brisbane demonstrated how co-design and speculative design structure new design approaches that authorise tacit knowledge and the redesign of design by lay experts.

Online debate among students on the Experimental Design blog while the ECDC cultural probe and Energy Babble were on exhibition in Brisbane raised several frameworks in which the designs could be rethought. One student mentioned Georg Simmel and pointed to what he had to say about the factors determining the value of a commodity: factors aren't separate from one another, but depend on the person's perspective formed by what they see as their purpose in engaging with a design, the context in which they do so, and their cultural and educational background. She contended that these factors interweave with each other and can't be isolated in an experiment. In her view, cultural probes countered the mass surveillance of the internet. Cultural probes were described by one student as 'researching through design', where the open nature of the 'process did not aim to benefit or produce a specific outcome'. Another student ventured that the value of cultural probes is the autonomy they provide; in other words, the 'freedom they give to the participants' who in turn 'respond unpredictably'. Another claimed that the probes were inclusive, or in her words 'not excluding individuals', which prompted the response that cultural probes are 'an alternative to purely objective analysis'. As another pointed out, 'critically', the cultural probes 'gathered data which supplemented the end-product [the Babble] instead of directly leading to it'. This facet of cultural probes was deemed important

soon be shown in the US. As with the energy communities we built it for, the Babble

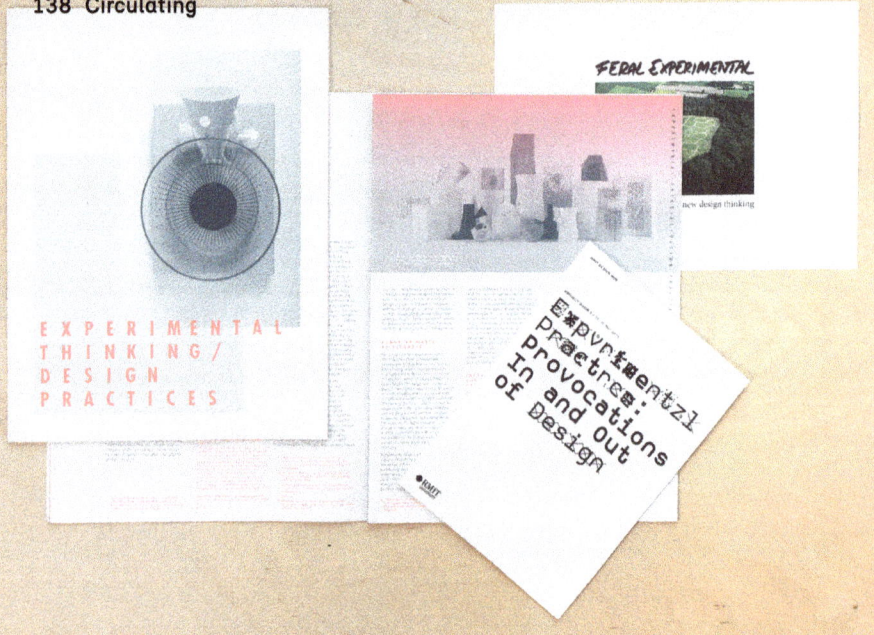

Katherine Moline invited us to show the Energy Babble at an exhibition she curated at the National Institute for Experimental Arts in Sydney. There, the Babble was framed as 'exploring the boundaries and intersections of experimental practice in design thinking, speculative design, participatory and co-design'.

given the context of the class discussion in which motivation and the crisis of agency was referred to as a cultural phenomenon that is closely connected to climate change.

A student's essay titled 'Consumerism, the Shift and Mass Customization' (2015), which was written while the work was on display in Brisbane framed experimental designs such as ECDC within critical theory, consumerism, and its imbrication in climate change. The author contended that the theme of control in consumerism is designed to create fantasies 'about fulfilling addictive desires'. Drawing on the writings of Arturo Escobar, she contrasted propositions that design has become open source and is therefore a positive force in globalisation with counter-arguments that framing design as an open process omits assumptions that are implicit in the concept of openness and neglects to account for the majority of the global population who have only limited access to digital technologies.

As co-curators Ahmed Ansari and Deepa Butoliya and I prepare the work for exhibition at Carnegie Mellon University in 'Climactic: Post

Normal Design' at Miller Gallery, I'm conscious that the ECDC cultural probes and Energy Babble will be reinterpreted in entirely new ways in the North American context (Carnegie Mellon University, 2016). Based on my experiences of discussing disparate interpretations of ECDC with co-curators, gallery visitors, and students in Sydney, Melbourne, and Brisbane, I return repeatedly to questions about cultural probes and autonomy, and the Energy Babble and surveillance. Most often, the Babble speaks to me as a Surrealist interruption to expectations about design. It is a non sequitur for Human-Computer-Interface debate because it refuses cognitive behaviour as a realistic reflection of how people engage with technology. Instead, the Energy Babble prioritises human irrationality and the unconscious as a data manifestor that distorts inputs and outputs. It baffles intentionality. Rather than see this as offensive, the real value of making the design's development open to the public in ECDC is that the data are available for all to interpret and reinterpret freely.

One question I continue to ask of the Energy Babble is whether its function is to mirror and intensify

the ordinary, everyday behaviours that must change to reduce energy consumption, despite the difficulty and the crisis of agency when challenged by the immensity of the scale of change ahead. As a design, it suggests that the future will depend on the capacity of humans – rather than of things – to adjust behaviours. In the ongoing conversations I've had with the Energy Babble over the past three years, as I've tried to make it work, and at times failed to do so, I think I've found its function in the era of fast fixes and instant gratification: things aren't prepared for the imminent climate change ahead. However, the open-source workbooks and images that the Interaction Research Studio have published about ECDC are a Surrealist Wunderkammer of affective design that tests its own efficacy in every city in which it has been exhibited in the southern hemisphere. As Herbert Read observed ninety years ago, the UK is the natural home of Surrealism, and it has much to tell us still about the affective dimension of reimagining the future (Herbert, cited in Hauser 2007: 15). 'Climactic: Post Normal Design',

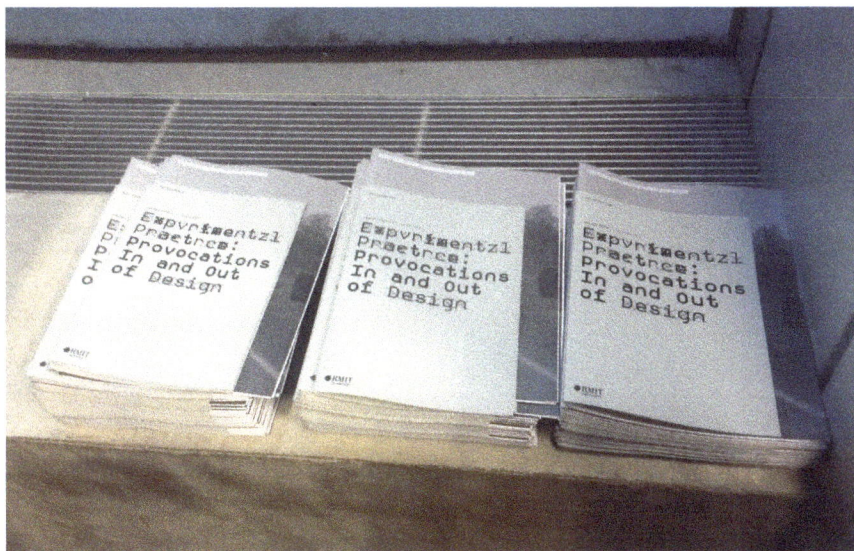

Another exhibition in Brisbane was curated by Katherine with Peter Hall and Beck Davis. This time, the intention was to 'emphasise embodied knowledge and further complicate questions about design in terms of curatorial and research practice, as well as making'.

References

Miller Gallery, Carnegie Mellon University, Pittsburgh, 4 November–11 December 2016 <http://millergallery. cfa.cmu.edu/exhibitions/ climacticpostnormal/>

'Experimental Practice: Provocations In and Out of Design', RMIT Design Hub, Melbourne, 11–29 May 2015 <http://designhub.rmit.edu.au/ exhibitions-programs/experimental-practice-provocations-in-and-out-of-design>

'Experimental Thinking/Design Practices', Griffith University Art Gallery, Brisbane, 18 September–7 November 2015 <www.griffith.edu.au/ visual-creative-arts/griffith-artworks/ exhibition-program/2015-exhibitions/ experimental-thinking>

'Feral Experimental: New Design Thinking', UNSW Galleries, Sydney, 18 July–30 August 2014 <www.niea.unsw.edu.au/events/ exhibition-feral-experimental>

Gaver, W., M. Michael, T. Kerridge, A. Wilkie, A. Boucher, L. Ovalle, and M. Plummer-Fernandez, 'Energy Babble: Mixing Environmentally-Oriented Internet Content to Engage Community Groups', *CHI 2015, Crossings*, Seoul, Korea, pp. 1115–24.

Hauser, K., *Shadow Sites: Photography, Archaeology, and the British Landscape 1927–1951* (Oxford: Oxford University Press, 2007).

Michael, M., '"What Are We Busy Doing?": Engaging the Idiot', *Science, Technology & Human Values*, 37(5) (2012): 528–554.

Moline, K., 'Bringing Experimental Thinking and Practice Together', in K. Moline and P. Hall, eds, *Experimental Thinking/Design Practices* (Brisbane: Griffith University Art Gallery, 2015a), pp. 8–9.

— 'Experimental Practice: Provocations', in *Experimental Practice: Provocations In And Out Of Design* (Melbourne: RMIT Design Hub, 2015b): 3.

— 'Finding Holes in Categorical Fences', in K. Moline and P. Hall, eds, *Experimental Thinking/Design Practices* (Brisbane: Griffith University Art Gallery/Art Works, 2015c): 4.

of interpretations as, variously, a new approach to environmental issues, an example of how

An updated version of the Feral Experimental show was curated in Melbourne. This time the Babble was shown alongside other design work as an example of 'new hybrid practices and collaborations are negotiating complex social and environmental challenges'.

design can operate to open new perspectives and engagements, and as both an outcome of and

The stuff of method: Open things and closed objects

Mike Michael

Introduction

This essay considers the role of objects in social scientific methodology. Of course, objects are necessarily part of conducting social scientific research. In methods such as interviews and focus groups, objects that make up recording equipment, the furniture and environment, utility and transportation systems render social science do-able (as well as on occasion subvert its feasibility – see Michael 2004). On top of this, objects can play a part in the crafting of data – as stimulus materials, they serve to prompt responses from participants, to trigger memories or focalise arguments, for instance. In ethnography, people's use of objects is key to understanding their cultural and social dynamics. Less common in social scientific research is the use of artefacts that are introduced by researchers into a particular setting familiar to participants who then live with the artefact for an extended period in the hope that it is 'open' enough to prompt reflection about some relevant issue or other. Of course, this is what the Energy Babble was designed to do, along with the other designs produced within the 'speculative design' tradition (see 'Design and science and technology studies' essay).

In this essay, then, attention is paid to how objects and things might be theorised as research tools. Drawing on various approaches, not least from science and technology studies (STS), objects and things are discussed in terms of their complex and heterogeneous constitution. Constitution in this case is understood in terms of 'eventuation': objects and things emerge in specific events that incorporate not only the design and content of the objects, but also what might otherwise be called its 'context'. Key here is the view that its precise constitution or composition arguably renders an artefact more or less closed (object-like) or open (thing-like) in the sense that it is more or less likely or able to specify how it can be used. Using this general framework,

the essay addresses how the Energy Babble was eventuated in a number of ways that combined open-ness and closed-ness in varying proportions. That is to say, we trace the ways in which what the energy babble 'is', and what it could yield as a methodological technology emerged out of a nexus of elements that were only recoverable in retrospect. As we shall see, this 'unknowability' of the Energy Babble was a reflection of our own affective relation to it and its prospective users.

In what follows, there is an initial introduction to some of the formulations of the object and the thing in STS. The particular approach adopted here is further developed through a discussion of the process philosophy of Stengers and Whitehead. Here, we see that object/things are admixtures of openness and closed-ness: we argue that things/objects are constitutively ambiguous. Nevertheless, the researcher, as a part of the eventuation of an object/thing, can serve to delimit the balance of openness and closed-ness, though here too we find ambiguity (and ambivalence).

Objects and things

In social science, objects have taken on an increasingly prominent position. Clearly, as novel artefacts and products of innovation, they have been studied in terms of their impact on society. Iconically, information and communication technologies are seen to shape society in a whole range of ways; globally, there is the putative emergence of virtual society or network society (e.g. Woolgar 2002); at the 'meso' level there are renewed forms of surveillance and audit that structure how organisations work (e.g. Bowker and Star1999; Power 1999); and at the microsocial level there is a reconfiguration of interpersonal relations and identities (exemplified in such figures as the 'calculated self' – e.g. Lupton 2015). However, mundane objects – clothes, paperclips, chairs – have also been subject to analysis. By virtue of the scripts they embody (Akich 1992; Latour 1991, 1992), or the propensities they encompass (Miller 2005), everyday artefacts are variously seen to afford and delimit, prescribe and proscribe particular practices. On this score, they are constitutive of social relations, just as they are themselves constituted through social processes. Thus, objects are quasi-subjects, part of the fabric of society, and, conversely, human subjects are quasi-objects (e.g. Serres 1982). And as Latour once put it: 'We are never faced with objects or social relations [...]. No-one has ever seen a social relation by itself [...] nor a technical relation' (1991: 110).

The ways in which objects affect people is particularly important here. To the extent that they incorporate particular scripts – sequences of actions that must be followed if the object is to 'work' – they impose a sort of morality, or a politics, even. There is a 'proper' way to do things, in other words. Yet, this is clearly discriminatory against people whose bodies might not operate in the ways presupposed by the scripts, or whose circumstances require different sorts of functions (Latour 1992). Thus, people can and do resist – or de-script – objects. What were seemingly 'closed' artefacts that function in specified ways and with specific requirements can be 'opened'. This opening can happen in numerous ways: through major political interventions such as the disability movement (Galis 2011); through local collective subversions (as in community re-purposing of technologies – De Laet and Mol 2000); or in individual reactions (for example, asking for immediate help to negotiate a recalcitrant technology – Michael 2000).

The idea of 'opening' the artefact, of bringing out possibilities that have otherwise been 'obscured' echoes a distinction made by Hans-Georg Rheinberger between technical object and epistemic thing. In his study of experimental practice, Rheinberger (1997) provides an account of scientists who, in building an experimental system, deploy stabilised elements – technical objects – such as various bits of equipment and types of materials. It is the juxtaposition and interaction of these elements that yields the uncertainty that scientists aim for in their pursuit of experimental knowledge. Out of this uncertainty emerge what Rheinberger calls 'epistemic things', which might be chemical reactions, physical structures, or biological functions that 'present themselves in a characteristic irreducible vagueness [... because they] embody what one does not yet know' (ibid.: 28). In other words, these 'things' are open, not yet fully disclosed, emergent through the uncertainties of the experimental system.

In light of this, we might want to disaggregate stuff into two categories – those that are 'closed' (objects), and those that are 'open' (things). However, as we shall see, things are rather more complex than this dichotomy suggests. To explore this in more detail, we need to turn to the processual works of Whitehead and Stengers (also see 'Design and science & technology studies' essay).

Process, and primary and secondary qualities

For Whitehead (1929, 1933; Halewood 2011) objects and things are 'actual entities' composed (concresced) out of multiple and heterogeneous components (prehensions) that can span such dichotomies as social and natural, organic and inorganic, micro and macro, conscious and the unconscious. One implication of this is that we need to attend to the specificity of this composition and the resulting emergent entity – to how the particular prehensions concresce in a particular event. In other words, what something 'is' depends on the discrete event of which it is a part, and through which it is eventuated. This means that we cannot assume that stuff has some sort of essence, that is, primary qualities, to which are added other secondary qualities. Thus, this perspective does assume a pre-existing entity such as a 'car' or a 'radio' to which secondary qualities are added, such as 'red' or 'retro'. What is eventuated is a 'red car' and a 'retro radio'. Or if there is a radio that is essentialised, or abstracted, this is also eventuated in its specificity by, say, a philosopher or an electrical engineer or a media executive: there is philosopher X's or engineer Y's version of the abstracted car.

If we accept that 'eventuation' entails the collapse of the distinction between primary and secondary qualities, then it follows that openness and closed-ness are not qualities that can be added to this or that artefact, but eventuate with the artefact in the specific process of that artefact's emergence. The question becomes: how is stuff rendered relatively more or less open or closed in its eventuation?

To address this question, we need to examine in more detail what goes into eventuation. In science and technology studies, it has long been known that a technical artefact cannot be dissociated from a panoply of constitutive elements. For present purposes, we can point to: the range of formal regulations, standardised components, and operating instructions (contained in manuals, for instance); the advertisements and publicity about benefits and advantages; accompanying narratives about actual functions and de facto workarounds; dramatisations of social as well as practical problems that attach to an artefact (e.g. Pfaffenberger 1992). This suggests that artefacts are routinely ambiguous and, as such, they evoke, in principle, a range of contrasting responses.

Having noted all this, in the course of everyday life, artefacts do not display this ambiguity. We can explore this further by referencing ideas from Gibson's ecological psychology theory of affordance (e.g. 1979; also, Ingold 1992): what is afforded to persons (and indeed to nonhuman animals) is indissolubly linked to the particular embodiment and ongoing actions of that person. So, what an object can 'do' is influenced by what the body of the person is capable of: a beam does not afford walking across if someone has no sense of balance. Needless to say, this can be altered, not least by 'enhancing' the body either socially (recruiting other people to help) or technologically (employing soles with more grip, using a balancing pole), or some combination of

to talk about the strange experience of developing, building, and deploying the Babble –

the two. Here, other affordances kick in – the affordances of poles, soles, and people: what we are usually witness to are cascades of affordance (Michael 2000). Further, what an object can do is affected by the person's unfolding 'plans' as they take shape and adapt in the flow of action: a table affords working on, or when one detects an earthquake, shelter.

To speak of the eventuation of an entity, in this context, is also to address the emergence of affordance. An entity is partly constituted by the relations entailed in the uses to which it can be put. Therefore, to reiterate, primary and secondary qualities are fused insofar as what an entity 'is' emerges out of its nexus of relations – a nexus that takes in bodies and their capacities, other entities and persons, plans and intentions, environmental and social occurrences and conditions. Any and all of these can conspire to 'fix' an entity, to occasion something's closed-ness – albeit just for that event. However (as seen in the 'Design and science & technology studies' essay), eventuation can also be understood as processual, as unfolding towards the not-as-yet. To clarify, while the elements of events concresce and mutually shape one another, what an event 'is' might become open, even though the stuff that eventuates through that event might nevertheless emerge 'closed': the point is that this closed-ness will vary depending on how the elements of an event concresce. As we shall detail below, even artefacts designed to be open – intended, as it were, to lure their users into a sense of ambivalence, playfulness, reflection and speculation – can end up occasioning users' closing-down of the artefact (as in, for instance, an instrumental failure – see Gaver et al. 2007, 2009; Michael 2016).

In the next sections, we take up the issues discussed above and explore them in relation to speculative design. In particular, we look at the work that goes into attempting to render an artefact open – not only through the process of design, but also through the overt and tacit procedures entailed in that artefact's implementation. As we shall see, a number of ironies are in operation: for instance, in order to make an entity open, it must also, in various ways, be closed.

Open/closed: Bodies/plans/stuff

As detailed elsewhere in this volume, the Energy Babble essentially comprises a combination of devices (individual Babbles) that are networked through a server-based system. By generating and distributing more or less comprehensible, energy-related statements derived from user input and various online sources, the Energy Babble was a research device that was designed to 'open up' the ways in which local energy communities understand, draw on, problematise, and undermine the issues that surround energy use and energy-demand reduction. The Energy Babble aimed to explore how community users dealt with energy matters as they were manifested in the news and in policy, in everyday practice within and across communities, and in individual and collective projects and aspirations. Informing the Babble research was the view that such engagements with energy

(and energy-demand reduction) are emergent, unfolding, immanent. The Babble, in all its idiosyncrasy, playfulness, and opaqueness, was designed to enable users to access to the potentialities of energy-demand reduction. By 'distorting' or 'ambiguating' energy-related information and its flow, and by broadcasting its semi-sensical statements in unpredictable ways, the Babble ideally should have provoked, prompted, and invited an openness to energy-demand reduction and its associated issues. What counts as a community? What comprises energy? What constitutes information? These are some of the sorts of questions that might have been inspired by the Babble's interjections (though ideally we would have preferred questions that we did not foresee).

But let us step back for a moment and ask: what needs to be in place for the Babble to operate in this way, as something that, in its openness, invites openness? If a central aim of 'speculative design' is to make such 'open' devices, what guarantee is there that they will work in this way? In short, there is no guarantee. As Gaver and colleagues have documented, devices can and do fail. The example of the Home Health Monitor (Gaver et al. 2009) is instructive in this regard. Designed to 'provide an intriguing reflection on the household's "mood"' (n.p.), it failed, prompting instead a series of critical responses. The Health Home Monitor used a series of sensors that measured such things as whether a sofa had been sat on, or a door opened, to generate a sense of the 'health' of a household displayed through such genres as aphorisms, pie charts, and photographs. These were intended to be ambiguous and playful so as to lure the occupants of the home into further reflection. Instead, the occupants 'instrumentalised' the system, criticising in terms of the accuracy of its output, or its lack of obvious utility. Gaver et al. put this lack of engagement down to a number of factors concerning the Home Health Monitor itself (for instance, the outputs were insufficiently meaningful, the outputs were also too thin when compared against the complexity of the system design). But they also point to the users themselves: they were not, it turned out, especially interested, neglecting, for example, to set the system within household routines. Reframing the Home Health Monitor in the terms of eventuation presented above, we might say that its designed affordances did not resonate with the plans and capacities of the householders. Indeed, we might say that the householders were relatively resistant to the device – interpreting it instrumentally rather than using it exploratively – and closing it, and their engagement with it, down, rather than opening it up towards not-as-yet engagements. Yet, this resistance might have been resourced by the design itself, including the way it was presented to the users, as reflecting the designers' interest in 'home health' (this stood in contrast to the deliberate reticence of the designers when they installed a previous iteration call the Home Health Horoscope – see Gaver et al. 2007).

The point here is that all these factors – these prehensions, we could say – combine or concresce in the specific eventuation of the speculative device. How they combine in the process is what is important. In this eventuation, quandaries are in evidence. The design must be an open thing – adequately ambiguous, opaque, and playful so as to enable speculative engagement. However, if it is too open it might fail to make sense, provoking unease, suspicion, or antipathy on the part of potential users. In other words, it becomes a closed object despite its planned openness. Conversely, users must be primed in ways that link their engagement with the purpose of the device (and the designers' broad research agenda of promoting openness). The potential threats posed by the device as an open thing – in particular, its possible lack of meaning – must be waylaid by the designers by adding meaning to it. However, to make the device too interpretable risks, ironically, shutting down its openness: it becomes a closed object. Succinctly, a fine balance must be struck between making the device too interpretable and not making it interpretable enough – too much of an open thing (and hence threatening), and too much of a closed object (and hence failing in its promise). The themes of promise and threat will be developed in the next section when we discuss how the openness and closed-ness of the Energy Babble was enacted.

The threat and promise of the Energy Babble

The ECDC team interacted with the energy communities in various ways and at various times before the Energy Babble was presented to them. At an initial presentation to representatives of the energy communities, through site visits by members of the teams to the communities, through the probe workshop, through the distribution and collection of the probe packs, through the ECDC website – all these occasioned opportunities to affect the relationship between the ECDC project and the communities themselves. Or, to put this another way, these were moments when the team could impact on the sort of reception their speculative device would receive on implementation. By the same token, the experience of the ECDC team with the various communities (which, as we have seen, was characterised by very different circumstances and divergent priorities) affected the ways in which the design process and implementation phase proceeded.

Of course, these emerging relations between ECDC and the communities operated at numerous levels. Initially, the relationship might be thought of as fairly 'abstract'. That is to say, in early encounters, the ECDC approach would have been viewed as a novel – indeed entertaining – approach to thinking about the study of energy-demand reduction. As one of the most researched constituencies in the UK, the energy communities could contrast the prospect of engaging with ECDC against the usual forms of social scientific investigation – interviews, focus groups, ethnography (see Clark 2008). What ECDC promised at a general level was an empirical process that was intriguing, strange, exciting, even. However, this must also be placed in relation to the exigencies of working within and for an energy community.

These communities were involved in a constant struggle to find funding, to raise their profile, to develop projects, and so on. Contrasted against the limited and shrinking resources of the communities, the seemingly generously funded social science projects of the Energies and Communities initiative were regarded – and this was made plain to us on several occasions – as a mis-use, if not outright squandering, of precious funds and resources. Why resource the study of energy communities when the government could be actually financially supporting those communities? Within this perspective, the ECDC project with its outwardly vague research agenda (initially there were only general principles and approaches in place rather than discrete research tools) must also have been regarded as, at the very least, a risky project. In summary, we can say that the project itself is seen both in terms of 'promise' (it will be something intriguing, entertaining, revelatory) and 'threat' (it is something that takes resources away from us, it will waste our time).

As we got closer to the deployment of the Energy Babble, the ambiguity of the project – its simultaneous promise and threat – became intensified, impacting on the ways in which the team went about installing (socially as well as technically) the Babble within the communities. We present details of the process of implementation and the communities' responses to the Babble elsewhere in this volume. Suffice to say that here we draw out some of the ways in which we went about the process of 'balancing' the promise and threat of the Babble (see also Wilkie and Michael, in press).

As we note in the essay on 'Design and science & technology studies', all social science research should be understood in terms of a method assemblage (Law 2004) in which the engagement between researchers and their 'objects of study' takes place in multiple ways wherein there is mutual shaping. On this score, we can note that in implementing the Babble there was tacit concern on our part as the ostensible researchers that our object was so strange that it would be seen as threatening, in the sense that it did not make sense to the prospective users. There were other responses we were worried about, too – that it would be dismissed as trivial (say, in terms of its output), or condemned as wasteful (in terms of the resources that went into its design and production). In light of this background hum of anxiety about its reception (which we might say reflected the ways in which we were shaped by the communities as we came to understand them), there was a temptation to diffuse the speculative aspects of the Babble: to downplay its oddness in favour of its potential utility, its usefulness. This is especially apparent in our convoluted efforts to navigate the dual elements of the speculative and the instrumental, the playful and the utilitarian, in the way we portrayed the Babble in a feature that appeared in a newspaper local to one of the energy communities (November 2013 issue of *Reepham Life* – see Figure 1).

To reiterate, taken as an actual entity that eventuates out of the combination of a nexus of heterogeneous elements,

impulses – and how to understand its reception by the energy communities. Was it a success?

Energy Babble will help Reepham reduce energy

REEPHAM is one of the UK's first towns to trial a prototype of new technology designed to engage communities in reducing energy consumption. The project addresses how to achieve an 80% reduction of the country's carbon emissions by 2050.

The "Energy Babble" has already been installed at both schools in Reepham, and units are planned for other venues in the town.

Designed by Professor Bill Gaver and his project team at Goldsmiths, University of London, the Energy Babble collects information relating to energy issues from an extensive network. A server process aggregates and transforms input sources into audio files that are broadcast by each device.

The idea is to open and promote constructive debate and involvement on energy reduction issues. Anyone with access to a Babble can send information to it, either verbally using the microphone or remotely via texting or the internet. The information is then logged, checked, stored and broadcast within a few minutes.

Goldsmiths was awarded a £795,000 grant to fund the Energy Babble over a three and a half year period.

Prof. Gaver said: "The most fundamental achievements will be the Babble on the one hand, and people's engagements with it and reactions to it on the other.

"The Babble itself explores technical

Matthew Plummer-Fernandez of Goldsmiths, University of London, with one of four Energy Babbles being installed in Reepham

possibilities, but also summarises, in a sense, the situation of people who are trying to make progress on environmental action at a community level.

"The reactions don't just tell us whether people like the Babble, but also – we hope – will reveal a lot of their knowledge, concerns and beliefs about environmental issues, as the many different ways people minimise their energy consumption."

He continued: "The Babble is not a blank slate like Twitter and other social media, but is far more centred on environmental issues. Also, because the Bab-

ble is a physical device it has a presence, and a social one, that on-screen systems like Twitter, etc., may not have.

"Using audio output means it is more pervasive; you don't have to look at it to engage with it. It is designed to highlight the ways people talk about environmental issues, and that is more or less all it does, all the time."

As a winner of the Low Carbon Communities Challenge, the Reepham Green Team were introduced to Goldsmiths two years ago to help develop the Energy Babble.

In total, 35 prototypes have been made available to trial nationwide, of which four units have come to Reepham, two of which are already installed in the schools.

Judy Holland

■ The Green Team would like *Reepham Life* readers' help in deciding where the other units should be installed. Please send your suggestions to: info@reephamlife.co.uk, or in writing to Reepham Community Press, Homerton House, 74 Cawston Road, Reepham, Norfolk NR10 4LT, or left at Very Nice Things in the Market Place.

www.reephamchallenge.org

Figure 1: From the November 2013 issue of *Reepham Life*. The Energy Babble being displayed by project member Matthew Plummer-Fernandez.

the Babble is enacted as an ambiguous mix of promise (it is interesting, engaging, playful, exciting) and threat (it is nonsensical, unapproachable, alienating). This, of course, resonates with the figure of the idiot discussed in the 'Design and science and technology studies' essay. As we note there, the idiocy of speculative design resides in the nonsensical-ness of its artefacts. Yet, at the same time, this nonsensical-ness is expected to inspire openness, rather than result in the closed-ness of incomprehension or antagonism. The argument here is that the idiocy of the Babble has to be rendered safe, unthreatening, engaging. After all, idiots can be intimidating and menacing as well as charming and intriguing. Our representations of the Babble as having 'utility' are, then, an attempt to reassure users, to diffuse its potential threat as incomprehensible, etc.

However, this account of the promise and threat of the Babble is perhaps overly 'cognitive' – it focuses on the explicitly articulable elements of the Babble and the sorts of relations it might form with its users. There are other less accessible elements that contribute to what the Babble 'is'. As we noted in the discussion of the processes of concrescence and eventuation, the elements that comprise an object are unconscious as well conscious, affective as

well as cognitive. To elaborate, an object is affective – it eventuates through its relations to the body, emotions, and the senses of the user. For some authors at least, such affective relations bypass, or sit parallel to, the conscious ideas and understandings of users (e.g. Massumi 2002; see Weatherall 2012). These affective dimensions operate in numerous ways and we highlight two in the case of the Energy Babble.

The first concerns the emotional connections built up between the ECDC team and the communities over time. To be sure, these are manifold and vary in their intensities. To start with, our initial site visits, the probe workshops, and the email updates (not least about the ECDC website) all contributed to shaping and sustaining particular affective relations between us and the energy communities. This need not be purely positive, of course: on the negative side, our work might be felt as a profligate waste of scarce resources. By contrast, energy community members might simply feel that we were 'nice people' showing interest in their efforts, predisposing them towards at least a tolerant or receptive, if not out and out welcoming, response. On this score, there is a degree of identification between designers and communities.

failure? A successful failure? This book is our attempt both to share the process of coming

Secondly, there was the aesthetics embodied by the Babble itself. The Interaction Research Studio is well known for its design and production of highly finished artefacts that have a certain beauty to them, and which are aesthetically intriguing, not least because they reference all manner of other more or less recognisable technologies and objects. In a sense, then, these designs affectively 'draw the user in' because they are suggestive of familiar technologies while also disturbing that familiarity, adding novelty. As should be evident, the design of the Babble references both scientific glass equipment and old-fashioned stand-up (candlestick) telephones. But this referencing would not have the same aesthetic impact if it were not 'beautiful'; that is to say, its elements form a well-proportioned 'coherent whole', and the 'finish' is highly crafted so that it appears 'well made' and 'careful'. The aesthetic affectivity of the Babble no doubt operates on many levels – in terms of the 'sensory affects' of beauty and finish as we have seen, but also in terms of 'social affects' in that the Babble displays care and craft, and thus a 'respect' for its prospective users. The point is that there are various 'aesthetic' elements of the Babble that together serve to be 'affective', 'drawing in' relations with users that are less threatening, more promising.

Concluding remarks: towards a notion of com-promise
In this essay, we have considered the ways in which entities are heterogeneously eventuated to be more or less indeterminate, as closed objects or open things. We have suggested that this is no easy dichotomy – that the seemingly closed are open, and vice versa. In the example of the Energy Babble this convolution of openness and closed-ness is particularly acute, and we traced a number of ways in which openness could precipitate closed-ness; that is, the open speculative character of the Babble might, for various reasons, prompt a reaction in which it were closed down. We tried to articulate these convolutions through the motifs of threat and promise. The Babble was simultaneously threatening (through, for instance, its specific incomprehensibility) and promising (through, for instance, the ways in which it was intriguing). Part and parcel of this were our own efforts to diffuse its potential to threaten by placing emphasis on its promise to fulfil instrumental or useful function. Ironically, this utilitarian promise threatens to derail the prospect of openness in the Babble and thus its speculative promise. In compromising on the openness of the Babble (closing it down through stressing its functionality), we compromise the Babble's openness (its capacity to engage the user in such a way as to explore the various meanings and enactments of energy-demand reduction, information, community, etc).

However, this casts a rather negative light on compromise. Perhaps there is something more interesting to be said about the notion of compromise as a way of coming to understand both speculative devices, but also objects/things more generally.

For our purposes, and etymologically speaking, we can think of 'promise' in terms of 'putting forward'. The specific 'putting forward' of the Babble entails a lure – the invitation to engage with its 'intrigue' (its playfulness, opacity, ambiguity). However, also being 'put forward' is a certain riskiness – the Babble might be incomprehensible, wasteful, trivial: as much as it lures, it can repel. As we have seen, there are various ways of deflecting antipathy – discursive (articulation through the language of functionality), social (the enactment of forms of identification), aesthetic (making the Babble 'beautiful'), and, of course, combinations of these (the 'care' embodied in the Babble's aesthetic crafting that signals 'respect' for the energy communities). However, we might also regard the members of the energy communities as themselves holding 'promise' – putting themselves forward in ways that engage with the Babble speculatively, of being open to the Babble's openness. Of course, as we have noted, promise can, from the perspective of the designers, additionally be seen as negative: participants can always 'put forward' resistance to a speculative device or enact reticence in their engagement with it.

The implication is that just as the Energy Babble 'puts forward', so too do its users. There is, one might say, a mutual promising: a 'putting forward together'. Given that together can be etymologically translated as 'com', this putting forward together can be grasped as a 'com-promise'. Accordingly, we would hope that com-promise does not possess the connotations of dilution or modulation that attach to certain versions of compromise. Instead, com-promise should evoke the complexity and convolution – indeed, involution (e.g. Ansell Pearson 1999) – of connections amongst the elements involved in eventuation. In the case of the Babble, com-promise is necessary for successful speculative research. The Babble being understood as a heterogeneously enacted entity, what it 'is' emerges from the multiple, varied, and shifting relations entailed in the actions of designers and users. The corollary point is that these designers and users are themselves composed of multiple, varied, and shifting relations. Out of this nexus emerges – hopefully – user engagement with the Babble that is speculative insofar as it begins to open up interesting questions about what counts as 'community', 'information', 'energy', 'environmentalism', and so on. We say 'hopefully' because, as we have hinted in the foregoing, what has entered into the process of com-promise only becomes apparent retrospectively. There is no way of guaranteeing that the nexus of connections that makes up a com-promise will 'work'. Nevertheless, we would suggest that the notion of com-promise holds a broader heuristic promise – that of illuminating how any device 'works' by alerting us to the involutions of promise and threat, closed-ness and openness, object-ness and thing-ness entailed in its eventuation.

References

Akrich, M., 'The De-scription of Technical Objects', in W.E. Bijker, and J. Law, eds, *Shaping Technology/ Building Society* (Cambridge, MA: MIT Press, 1992), pp. 205-24.

Ansell Pearson, K., *Germinal Life* (London: Routledge, 1999).

Bowker, G.C., and S.L. Star, *Sorting Things Out: Classification and its Consequences* (Cambridge, MA: MIT Press, 1999).

Clark. T., '"We're Over-Researched Here!" Exploring Accounts of Research Fatigue within Qualitative Research Engagements', *Sociology*, 42(5) (2008): 953-70.

de Laet, M., and A. Mol, 'The Zimbabwe Bush Pump: Mechanics of a Fluid Technology', *Social Studies of Science*, 30 (2000): 225-63.

Galis, V., 'Enacting Disability: How Can Science and Technology Studies Inform Disability Studies?', *Disability & Society*, 26(7) (2011): 825-38.

Gaver, W., J. Bowers, T. Kerridge, A. Boucher, and N. Jarvis, (2009). 'Anatomy of a Failure: How We Knew when our Design Went Wrong, and What We Learned from it', Paper Presented at *Conference on Human Factors in Computing Systems*, Boston, MA, 2009.

Gaver, W., P. Sengers, T. Kerridge, J. Kaye, and J. Bowers, 'Enhancing Ubiquitous Computing with User Interpretation: Field Testing the Home Health Horoscope', *Proceedings of CHI ACM*, 2007, pp. 537-46.

Gibson, J. J., *The Ecological Approach to Visual Perception* (Boston, MA: Houghton Mifflin, 1979).

Halewood, M., *Alfred North Whitehead and Social Theory: The Body, Abstraction, Process* (London: Anthem Press, 2011).

Ingold, T., 'Culture and the Perception of the Environment', in E. Croll, and D. Parkin, eds, *Bush Base: Forest Farm – Culture, Environment and Development* (London: Routledge, 1992).

Latour, B., 'Technology is Society Made Durable', in J. Law, ed., *A Sociology of Monsters* (London: Routledge, 1991), pp. 103-31.

— 'Where are the Missing Masses? A Sociology of a Few Mundane Artifacts', in W.E. Bijker, and J. Law, eds, *Shaping Technology/Building Society* (Cambridge, MA: MIT Press, 1992), pp. 225-58.

Lupton, D., *Digital Sociology* (London: Routledge, 2015).

Massumi, B., *Parables of the Virtual* (Durham, NC: Duke University Press, 2002).

Michael, M., *Reconnecting Culture, Technology and Nature: From Society to Heterogeneity* (London: Routledge, 2000).

— 'On Making Data Social: Heterogeneity in Sociological Practice' *Qualitative Research*, 4(1) (2004): 5-23.

— 'Speculative Design and Digital Materialities: Idiocy, Threat and Compromise', in E. Ardevol, S. Pink, and D. Lanzeni, eds, *Designing Digital Materialities: Knowing, Intervention and Making* (London: Bloomsbury, 2016), pp. 99-113.

Miller, D., Introduction in S. Küchler, and D. Miller, eds, *Clothing as Material Culture* (Oxford: Berg, 2005), pp. 1-19.

Pfaffenberger, B., 'Technological Dramas', *Science, Technology and Human Values*, 17 (1992): 282-312.

Power, M., *The Audit Society* (Oxford: Oxford University Press, 1999).

Rheinberger, H-J., *Toward a History of Epistemic Things: Synthesizing Proteins in the Test Tube* (Palo Alto, CA: Stanford University Press, 1997).

Wetherell, M., *Affect and Emotion: A New Social Science Understanding* (London: Sage, 2012).

Whitehead, A. N., *Process and Reality. An Essay in Cosmology* (New York: The Free Press, 1929).

— *Adventures of Ideas* (Cambridge: Cambridge University Press, 1933).

Wilkie, A., and M. Michael, 'Designing and Doing: Enacting Energy-and-Community', in N. Marres, et al., eds, *Inventing the Social* (Manchester: Mattering Press, forthcoming).

Woolgar, S., ed., *Virtual Society? Technology, Cyberbole, Reality* (Oxford: Oxford University Press, 2002).

Like the project itself, none of this is quite settled yet. Nonetheless, we hope, dear

Afterword
Bill Gaver

So what have we learned from the Energy Babble and, more generally, the ECDC project? It may seem flippant, but it's tempting to conclude that 'life is complicated, and so was this project'.

Of course, we could have made it simpler. We could have focused on developing a tool that would highlight and address a primary issue faced by the communities. A 'better' energy-demand meter, for instance. Or a forum for sharing information about government policy. Or a tool for measuring and trading best practices for engaging the wider community. Focusing our investigations and design work on any of these topics would have made it much easier to claim an understanding of what the most important issues are for the energy communities, and to claim that our design work had successfully addressed those issues.

Narrowing our focus to one or a few issues, however, would have done an injustice to what we actually found in the communities. Instead, through our visits and conversations it became clear that all these issues, and many, many more, were deeply intertwined to produce the situations in which the communities acted, and that these issues could be seen through a diversity of perspectives ranging from public engagement with science, to speculation, to studio culture, to the choice of bot architecture.

Choosing to keep our focus broad, and to ensure that our design and thinking matched the breadth of what we saw, and to avoid offering simple explanations or solutions is what made the project complicated – as complicated as the situation we addressed. To a great degree, our ability to appreciate and work with that complication (rather than pulling at a loose thread that we could claim as the most important issue) reflects the engagement of design with STS. The concept of entanglement itself, along with notions of assemblage, performativity, and conceptual figures such as the idiot all informed and enriched the design work described in these pages.

The Energy Babble reflects the complexities of the communities' situations, and does little, if anything, to resolve them. What is the use in this? We would suggest that, as the lengthy discussions with community members – discussions sparked by their encounters with the Babbles – indicated, attempts to 'solve' energy consumption, whether through policy, technology, or community organisation, seem to be always impeded by the many other factors they overlook. This is not to undermine the many successful and downright heroic activities pursued by the communities. Nonetheless, in talking to them it becomes clear that attempts to identify the 'most important' issues are chimerical – that pulling on a thread that appears to be loose inevitably brings the whole tangle along with it. Clearly, the Babble's insistence not only on reflecting but alson condensing and amplifying that tangle may seem frustrating compared with endeavours that claim to identify and solve 'most important issues'. In its stubborn insistence that the tangle of issues faced by the energy communities is real, present, and impossible to ignore, however, the Babble may paradoxically represent the more effective way forward.

reader, that you have enjoyed our curious tale.

Mattering Press titles